The
Arid
Zones

UNIVERSITY LIBRARY OF GEOGRAPHY

Edited by W. G. East
Professor of Geography
University of London

Kenneth Walton

The Arid Zones

Routledge
Taylor & Francis Group

LONDON AND NEW YORK

First published 2007 by Transaction Publishers

2 Park Square, Milton Park, Abingdon, Oxfordshire OX14 4RN
711 Third Avenue, New York, NY 10017

Routledge is an imprint of the Taylor & Francis Group, an informa business

First issued in hardback 2017

Copyright © 2007 Taylor & Francis

Library of Congress Catalog Number: 2006051885

Library of Congress Cataloging-in-Publication Data

Walton, Kenneth, 1923-
    The Arid Zones / Kenneth Walton.
        p. cm.
    Originally published: Chicago : Aldine Pub. Co, 1969.
    Includes bibliographical references (p.).
    ISBN 978-0-202-30928-6 (alk. paper)
        1. Arid regions. I. Title.

GB611.W3   2007 551.41'5—dc22   2006051885

ISBN 13: 978-0-202-30928-6 (pbk)
ISBN 13: 978-1-138-53428-5 (hbk)

CONTENTS

FIGURES

I

THE NATURE AND CAUSES OF ARIDITY

Fundamental to the creation of the characteristics of the arid environment is the climate. This conditions the way in which landforms, vegetation, animals, soils and modes of life differ in degree and kind from those of the humid areas of the earth's surface. From the interior deserts of Central Asia and the Sahara to the foggy but arid coastlands of Peru or South-west Africa scarcity of water in the upper layers of the rocks and the surface deposits is a limiting factor in land use and development. Since water shortage is often caused by low rainfall linked with high evaporation rates, it is logical to examine the extent and causes of the low precipitation of the dry lands which cover about one-third of the earth's surface. Unfortunately through the nature of the environment population densities are normally low and accurate analyses of the climate can be based only on recently established scientific stations and meteorological observatories such as those established by the French in the Sahara and by the Chinese and Russians in Central Asia. More climatic statistics are available for the Algerian Sahara and the American deserts than for the Libyan desert, but records rarely extend for fifty years and stations are often very far apart. Where they have existed for a long period they have not always recorded the information which the climatologist and biogeographer require. Climatic maps suffer from the lack of these statistics since they tend to be constructed from the data of a few widely spaced stations with interpolated values for the intervening areas. It is apparent that the simple curves of desert isohyets (lines showing equal amounts of rainfall)

are broad generalisations which will be modified in years to
come.

Arid and semi-arid lands: definitions of aridity
This lack of adequate climatological information is, in part,
responsible for the numerous and often unsatisfactory attempts
to delimit the margins of the arid zone by climatic data, and to
divide the arid area into its more and less humid components.
Since aridity is primarily a function of rainfall, temperature and
evaporation, it is wrong to define it in terms of one parameter
alone, although average annual rainfall totals have been frequently
used as a simple index of aridity. Some workers have taken the
humid boundary as the 254 mm (10 in) isohyet and the margin of
the inner arid zone as the 127 mm (5 in) isohyet. The southern
margin of the Sahara was drawn where the annual rainfall is
250 mm (9·8 in) and the equatorial margin of the semi-arid
savanna at the 400 mm (15·7 in) isohyet. These are, however,
oversimplifications which ignore the influence of temperature on
the efficiency of precipitation but may be of value if the isohyet
chosen is related to a change in the character of the vegetation,
land-use or way of life. In this respect the 400 mm (15·7 in) isohyet,
said to define the southern limit of the North African arid zone,
is of significance. To the north of this isohyet agriculture cannot
take place without irrigation, the need for which is frequently used
to define the arid lands.

For geomorphological processes, vegetation and agriculture,
other climatic phenomena such as the season, duration and intensity of rainfall are as important as the total amount while
temperature, which in part controls the rate of evaporation, is
also of great significance. That evaporation exceeded precipitation
in the arid zone was recognised by Penck in 1910 when he de-
limited the boundary of the dry lands at the places where evaporation and precipitation are equal in amount. There is clearly an
important link between temperature, precipitation and evaporation but unfortunately while temperature and precipitation are
easy to measure and have been recorded at stations in the arid
lands for many years, the measurement of evaporation is more
difficult and records are both of shorter duration and much less
plentiful. This lack of data on evaporation prevented Penck's
concept from being developed on a large scale and many
indices of aridity and related classifications of climate have

been attempted using the criteria of rainfall and temperature alone.

Recognising the limitations of rainfall statistics, Köppen in 1918 linked temperature with rainfall to define the boundaries of the desert and steppe lands. In his scheme of climate classification the humid boundary of the deserts, in areas with rainfall well distributed throughout the year, is linked with the 200 mm (7·9 in) isohyet when the mean annual temperature is 5–10°C (41–50°F), but it follows the 320 mm (12·6 in) isohyet in areas with a mean annual temperature of 25°C (77°F). The corresponding figures for the humid margin of the steppe are 400 mm (15·7 in) at 5–10°C and 640 mm (25·2 in) at 25°C.

An elaboration of this scheme recognised that the efficiency of the rainfall varies according to the season of incidence; cold-season rainfall is more effective in areas with sufficiently high temperatures for plant growth since less is lost by direct evaporation than in the hot-season rainfall regimes. This is shown by the difference in agricultural potential of areas with the same rainfall on the Mediterranean and Sudanese margins of the Sahara. As an example, the margin of the humid steppe occurs at the 750 mm (29·5 in) isohyet in a summer rainfall regime with a mean annual temperature of 25°C (77°F). For an area with winter rainfall and the same mean annual temperature it corresponds to the 530 mm (21 in) isohyet. Köppen's amended climatic classification (the first was based on vegetation zones) is linked with these seasonal characteristics of temperature and rainfall. The steppe climate (BS) and the desert climate (BW) are further differentiated as 'h' (heiss) with mean annual temperatures of over 18°C (64·4°F) and 'k' (kalt) where the mean annual temperature is less than 18°C and the warmest month about 18°C, i.e. with a cold winter. Coastal deserts are distinguished by the suffix 'n' (nebelig) indicating frequent mist and fog.

Other indices of aridity used to prepare maps of humidity provinces derive from the work of Lang and de Martonne. They have been used to demarcate the dry lands of the USA and Australia. Lang produced the 'Rain Factor Index' obtained by dividing the mean annual precipitation in millimetres by the mean annual temperature in °C—the $\frac{P}{T}$ ratio $\left(\frac{P\ mm}{C°}\right)$. Regions with a $\frac{P}{T}$ ratio of under 40 were defined as arid. Yuma in Arizona has a

$\frac{P}{T}$ ratio as low as 3·5 while Gadames in Tripolitania has a value of only 0·7. Places with insignificant rainfall, such as Insalah in the Sahara, Assuan in Egypt and Walvis Bay in South-west Africa, have $\frac{P}{T}$ ratios of zero.

De Martonne slightly modified Lang's $\frac{P}{T}$ ratio to produce, in 1928, an index which also uses temperature and rainfall data— the Index of Aridity—$\left(\frac{P}{T+10}\right)$ where P represents mean annual precipitation expressed in millimetres and T=Temperature in degrees Centigrade. If this formula is applied, Yuma in Arizona has an index of 2·4 and Lima one of 2, so that both are within the limit of the true desert which de Martonne defined by an aridity index of 5. The dry steppe boundary, or the limit of cultivation without irrigation, was drawn where the aridity index had a value of 10 (cf. Teheran's index of 9½). In 1942 de Martonne amended the formula by including a representation of the total average rainfall (p in mm) and the average temperature (t in °C) of the driest month. The formula thus became $\dfrac{\dfrac{P}{T+10}+\dfrac{12\,p}{t+10}}{2}$ which gives values of less than 5 for the Sahara and Death Valley in California. Denver, Colorado, on the other hand, has an index of 18 and would thus not be considered to lie within the arid zone.

The formulae of Lang and de Martonne are both open to the criticism that they make evaporation appear to be a function of temperature alone although it is related to many factors including the amount of moisture in the soil, the soil type and texture, wind velocity, atmospheric pressure, relative humidity, plant cover and land-use. Nevertheless the definition of the arid lands provided by Lang and de Martonne is a reasonable approximation using only the means of the climatic parameters of temperature and rainfall. As with all the formulae based on averages, however, the assessment is confused by the high variability of rainfall from year to year which is a reality of the arid lands. Compare, for instance, de Martonne's index of aridity calculated for Yuma in 1899 with that for 1905; in 1899 Yuma received 25 mm (1 in) but in 1905 more than 280 mm (11 in). Rainfall variability of such magnitude

discourages unthinking reliance on even the more refined formulae devised to differentiate the arid from the humid lands and
degrees of aridity within the arid lands.

An attempt was made by Meyer in 1928 to overcome the
problem of the lack of information on actual rates of evaporation. He regarded evaporation as a function of the saturation
deficit which can be determined if the figures for temperature,
rainfall and relative humidity are available, i.e. by using the
moisture characteristics of the atmosphere at the recording
station. Using the absolute saturation deficit of the air obtained
by subtracting, for the prevailing air temperature, the actual from
the maximum vapour pressure possible at that temperature (the
latter is available from dew-point tables), he produced the Precipitation–Saturation deficit ratio (or the $\frac{P}{SD}$ ratio where P is
expressed in millimetres of precipitation and SD in millimetres of
mercury). On this basis the boundary of the semi-arid lands would
have a value of about 89 and that of the arid zone less than 44.

Although the $\frac{P}{SD}$ ratio does not take into account all the factors
controlling evaporation it is a more reliable index than those
which use only temperature and precipitation; unfortunately
although relative humidity values are more frequent than those
for evaporation they are still not readily available enough to
provide a practical solution to the problem. Nevertheless the most
accurate delimitations of the arid zone using climatic parameters
must use measured values of evaporation, a task which Thornthwaite in 1931 attempted with the formulation of the $\frac{P}{E}$ ratio or
Precipitation-Effectiveness Index.

The $\frac{P}{E}$ ratio is essentially an index of the efficiency of precipitation and attains its maximum value when the rate of evaporation from a free water surface is known. The index is obtained by
determining the sum of the $\frac{P}{E}$ ratios for each month of the year
and multiplying by 10 to eliminate fractions. $\left(\frac{P}{E} = \sum^{12} \times 10\right.$
where P = precipitation and E evaporation, both expressed in
inches.) For stations where evaporation data are not available,

Thornthwaite provided a formula based on mean monthly precipitation and temperature $\left(\dfrac{P}{E} = \sum^{12} 115\left(\dfrac{P}{T-10}\right)^{\frac{9}{10}}\right.$ where T is the mean monthly temperature value in °F). Compared with Meyer's values of 89 and 44 for the semi-arid and arid margins, the Thornthwaite values are 31 and 16 respectively.

In the former French Sahara there are about thirty meteorological stations which measure evaporation and Capot-Rey defined the margin of the Sahara using data from these records. The average of the ratios, annual rainfall/annual evaporation and rainfall/evaporation for the wettest month, $I = \dfrac{100\dfrac{P}{E} + 12\dfrac{p}{e}}{2}$, provides by this method an index of rainfall efficiency. For stations with the same annual rainfall amount the lowest index is obtained when the rainfall is in the warmer months. Values of between 4 and 5 indicate the desert limit. The success of this method depends on the frequency of stations which measure evaporation and, in consequence, the northern limit of the Sahara was more accurately defined than the southern.

If the amount of evaporation in arid areas is difficult to obtain, it is clear that the amount of transpiration by plants, an important means of water loss, will be even less readily available. Yet it has been shown by Thornthwaite that the potential evapo-transpiration values are a most accurate guide to the climatic differentiation of the arid lands. Potential evapo-transpiration is the amount of water which will return to the atmosphere from ground completely covered with vegetation where there is sufficient soil moisture for the use of vegetation at all times. That such conditions do not, by definition, obtain in the arid zone does not detract from the value of the method since it represents the water need of plants and cultivated crops and gives a useful guide to agricultural potential and to the amount of rainfall or irrigation required for a particular crop under the given climatic conditions. It does not, of course, take into account the problems imposed by the socio-economic systems of the farmers themselves. By this index, derived by Thornthwaite in 1948, the potential evapo-transpiration value for the semi-arid margin is 20 and for the arid margin 40.

Few stations have so far recorded evapo-transpiration values so Thornthwaite provided a formula using data from latitude and temperature alone.[1] Using this technique, Peveril Meigs in 1952 constructed for UNESCO (fig. 1) a map of the world's arid lands divided into semi-arid, arid and extremely-arid categories (the extremely-arid is defined as that which may have at least twelve consecutive months without recorded rainfall and where the rainfall lacks a seasonal rhythm). The table below indicates the area of the dry lands which fall into the three categories.

	Square miles
Semi-arid	8,202,000
Arid	8,418,000
Extremely-arid	2,244,000
Total	18,864,000

This represents about 36 per cent of the total land area of 52 million square miles. Maps on larger scales are now being constructed which will represent in more detail areas where drought conditions are characteristic.

It has been shown that there are many difficulties inherent in formulating climatic indices of aridity and some workers would prefer to use vegetation as an index. Provided there had been no interference by animals or man, which is doubtful, the vegetation would indicate the climatic values under similar edaphic (soil-forming) conditions. The vegetation of the arid zone is either tolerant to drought or drought avoiding or drought evading (see Chapter 4). It is often *xerophytic*, i.e. living in a comparatively dry habitat, and is frequently sparse and widely spaced with areas of bare soil between the patches of plant cover. Field surveys of vegetation in the arid zone are scarce although it is hoped that they will increase, while air photographs may speed up the work despite the difficulties imposed by the highly mobile character of steppe and desert boundaries. In Tunisia, for instance, in the four-year low rainfall period, 1944–7, the desert margin stood 270 km farther north than in the moister four-year period from 1931 to

[1] Khosla produced, in 1949, a simple formula for evapo-transpiration

$$Lm = \frac{Tm - 32}{9 \cdot 5}$$

where Lm is the monthly water loss or evapo-transpiration in inches, Tm is the mean monthly temperature in °F. When the Tm is less than 40°F Lm is as follows:

Tm	40°F	30°F	20°F	10°F	0°F
Lm	0·84″	0·70″	0·60″	0·50″	0·40″

Values obtained by this method are almost identical with those obtained from the more elaborate Thornthwaite computation.

1934. In 1947 even the olive trees at Sfax near the Mediterranean coast lost their leaves and the effect on the distribution of less deeply rooted vegetation was very marked. It is clear that such wide variations in the arid and semi-arid margins are to be expected in these zones of high rainfall variability; the boundaries are zones of transition rather than lines clearly demarcated by abrupt changes in species or associations. When the vegetation of the arid zones is defined by the rigid criterion of its mechanisms to combat drought, the area of the dry lands has been measured as follows:

	Square miles	
Semi-arid		
Sclerophyll brushland	1,180,000	
Thorn forest	340,000	
Short grass	1,200,000	
	———	2,720,000
Arid		
Desert grass savanna	2,300,000	
Desert grass, desert shrub	10,600,000	
	———	12,900,000
Extremely arid		
Desert	2,430,000	
	———	2,430,000
Total		18,050,000

This represents 35 per cent of the land area of the globe and is only 1 per cent less than the estimate of the arid lands measured on a climatic basis. There are, however, large variations in the areas assigned to the various categories which underline the problems of classification.

The relationship between the arid areas and the regions where the drainage of the rivers does not reach the sea has often been noted. De Martonne used the term *endoreism* to describe areas of interior drainage, a phenomenon employed by Richthofen to differentiate the arid regions of Central Asia from the more humid periphery through which flow the major rivers of southern and eastern Asia. According to de Martonne (fig. 1), the *areic* regions, whose drainage does not reach the sea, cover approximately 33 per cent of the land surface, a proportion which agrees well with the figures already given using climatic and vegetation indices. The

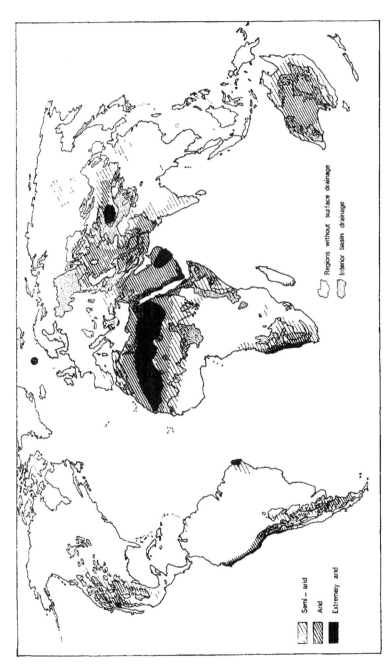

Fig. 1. Distribution of grades of aridity, basins of interior drainage and areas without regular surface run-off. (After Meigs, by courtesy of UNESCO, and de Martonne)

Semi – arid

Arid

Extremely arid

Regions without surface drainage

Interior basin drainage

boundaries of the basins of interior drainage conform broadly
with climate and vegetation boundaries and it might be possible
to differentiate the arid from the semi-arid by an analysis of the
frequency with which the watercourses actually carry water, if
seasonally regular flows, as from snow melt in the basins of
Central Asia, and sub-surface flows in the alluvium of river beds
are excluded. Yet whatever the subtle variations in the vitality of
the hydrological network, it is important to remember that for
one-third of the earth's surface the rivers do not reach the oceans;
such rivers are not linked to the general base level provided by the
sea and this has important consequences for the evolution of the
land forms of the arid zone.

The soils of the dry land regions possess distinguishing characteristics which permit a demarcation of the arid area although
complications arise from the climatic variations which have
occurred in the past. Arid zone soils are usually shallow, calcareous and only slightly weathered. These *pedocals* (p. 73) are
found wherever evaporation exceeds precipitation in both middle
and low latitudes. The climatic processes result in the deposition
of the soluble carbonates at the base of the moist layer, whose
depth below the surface varies with the degree of aridity and with
the degree of slope. It is near the surface in the very arid areas but
may be as much as three feet down on the more humid margins
where the *chernozems* (p. 74) show some slight calcification as
they change to the arid chestnut-brown soils where the grassland
becomes degraded with the lowering of $\frac{P}{E}$ values. There are, however, certain dangers in equating the *pedocals* with the arid areas
since it is also possible to find deeply weathered and leached soils in
desert and steppe lands. Such soils as those of the Alice Springs
area of Central Australia have been inherited from past climatic
and weathering cycles when the effective precipitation was higher.
If the pedocals are taken as characteristic of the arid zone, such
soils cover about 43 per cent of the land surface, which is nearly
10 per cent higher than estimates obtained by other methods. To
what extent this discrepancy is due to lack of precise information
on the areal distribution of the pedocals is unknown.

Enough has been said to indicate the problems and difficulties
of delimiting the arid and semi-arid areas. The fact remains that
for at least one-third of the land area of the globe lack of moisture
is the limiting factor for vegetation, animal life and land-use.
Plants, animals and human beings must adapt themselves to an

existence where they are constantly faced with the problem of water shortage, often under extremely high temperature regimes. In so doing, a complex physical/biological relationship is established, a dynamic relationship which betrays a constant adjustment to changes in precipitation and evaporation and makes the arid environment of vital interest. Most people think that they can recognise aridity when they see it but the history of human occupation of the arid lands indicates that a survey of indices of aridity is not of academic interest alone. They are essential for a true appreciation of variations in the arid environment.

The causes of aridity

Lack of precipitation in relation to the prevailing temperature and evapo-transpiration values is one of the principal factors in the creation of desert and semi-desert areas. Areas of low rainfall are often distant from the sea in the interior of continents, and the aridity is enhanced when high ground checks the entry of onshore winds. The arid lands frequently coincide with areas of permanently high atmospheric pressure although seasonally low-pressure zones may also lack precipitation under certain conditions. Some extremely arid areas are found, paradoxically, adjacent to the prime source of atmospheric moisture, the oceans. It is noteworthy also that the humid western margins of Eurasia give place to transitional arid types as where the Continental and Mediterranean climates of the Old World degenerate eastwards towards the heart of the continent.

Continentality and aridity are often coincident. Many regions of Central Asia have less than 254 mm (10 in) of rainfall which is quite ineffective when concentrated principally in the summer months when mean July temperatures often exceed 20–25°C (68–77°F). In winter the wind systems associated with the East Siberian anticyclone approach the dry lands of Central Asia from the north having crossed hundreds of miles of land. By the time they reach the dry interior they are low in humidity and high in evaporative power due to their passage from higher to lower latitudes, from colder to warmer areas. Only the Dzungarian desert, to the east of Lake Balkhash, receives some cyclonic rain from moist air transported from the Atlantic. When the East Siberian anticyclone disappears in summer, a low-pressure area is formed over the intensely heated interior of the continent while, to the west, an extension of the Azores High Pressure system develops across southern Europe and dry north-westerly winds prevail over the western parts of Central Asia. The eastern deserts, however, come under the influence of the restricted monsoon circulation and here summer aridity increases with distance from the Pacific Ocean, although the eastern part of the Takla-Makan betrays the influence of the monsoonal regime with its summer maximum of precipitation. It is, therefore, not only the absolute distance from the oceans that is important, but distance from the ocean from which come the moist airstreams.

The dryness of the air over Central Asia, however, is aggravated by the mountain barriers of the Tien Shan and the Pamirs in the west which prevent the ingress of moist Atlantic air except to Dzungaria. To the south the climatic barrier of the Himalayas eliminates the influence of warm moist monsoonal air from the Indian Ocean. It is only to the east that the Central Asian arid area is relatively open to winds drawn in from the Pacific Ocean so that the farther west one goes the greater the aridity and impoverishment of the vegetation.

In Australia the location of the Eastern Highlands athwart the direction of the prevailing south-easterly winds so accentuates the aridity of the continental interior as to produce an arid area of over 1,000,000 square miles. Adverse location of the principal relief features may not only enhance an aridity due to distance from the sea but may also deprive an area of precipitation which would be available from the planetary wind circulation. In the intermontane basins of the south-western USA, the aridity results from the rainshadow effect of the Sierra Nevada interposed in the path of the winter westerlies and the summer winds drawn in from the sea to fill the low pressure produced by the intense local heating of the land surface. The extremely arid areas with less than 127 mm (5 in) mean annual precipitation are found in the basin of the lower Colorado—Yuma has 84 mm (3·3 in). However, in Utah, although precipitation was much higher—Salt Lake City 1,331 m (4,366 ft) above sea-level has 406 mm (16 in)— the Mormons found it necessary in 1847 to site their farms by streams and to grow crops by irrigation. The mountain ranges which rear above the arid basins have slightly higher precipitation, making them valuable as a source of irrigation water for the waste-filled depressions at their base. In the temperate desert of Patagonia in South America, the principal cause of aridity is the presence of the Cordillera in the path of the prevailing westerlies since the mountains create a pronounced rainshadow on the leeward side.

The greatest deserts of the world, the classic zonal or hot deserts, are the result of the global distribution of pressure and winds. The extensive chain of arid lands from the Atlantic Ocean to north-west India and Pakistan straddles the Tropic of Cancer. In the southern hemisphere the deserts along the Tropic of Capricorn, apart from those in Australia, are restricted in size by the narrowness of the southern continents of Africa and South America. The Sahara and Libyan deserts, the deserts of Arabia, the Atacama of South America, the Kalahari of southern Africa and the Great Australian desert are all related to cells of high atmospheric pressure. In these high pressure zones of the 'horse latitudes', the upper air, moving polewards from the equatorial low-pressure belt, accumulates, as a result of the increasing in-fluence of the rotation of the earth, in a belt of westerly winds blowing along the parallels of latitude. Assisted by the gradual loss of heat through radiation as it moves towards higher latitudes from the equator, the air subsides on a large scale to produce areas of high atmospheric pressure.

Complications in the size and distribution of the high-pressure belts are introduced by the differential reaction to insolation of land and sea and the high-pressure belt breaks up into component cells. Because of the larger proportion of land in the northern hemisphere the high-pressure belt is less continuous than in the southern and is much more subject to seasonal variations. In winter, for instance, the Cancer high-pressure belts extend almost continuously from the eastern Pacific in an easterly direction to the coast of Asia; in summer, on the other hand, the intense heating of the continents introduces low-pressure systems over southern Asia and the south-western USA, although the latter is much smaller than its Asian counterpart. The Sahara is a shallow anticyclonic zone in the northern winter and an extension of the Azores High in the northern summer.

The winds which blow in the lower atmosphere from the high-pressure cells to the equatorial low pressures are the Trades. Traversing the hot deserts of the 'horse latitudes', they help to extend arid conditions to lower latitudes. They blow with seasonal regularity, little turbulence and at a modest speed. Coming over vast stretches of land the Trades are dry and desiccating; their moisture capacity is increased as they blow from cooler to warmer latitudes. In the Sahara they commence a few hours after sunrise and gradually subside in the evening so that the air is frequently still during the night. This is a partial explanation of the rapid

cooling of the air in contact with the ground after nightfall which
produces significant temperature inversions. High-pressure zones,
with subsiding air and dry surface winds, are areas unfavourable
to precipitation; they constitute the driest lands of the earth's
surface with low and highly irregular rainfall which increases in
quantity and seasonal regularity towards the equator and towards
the poles.

Aridity may also occur in seasonally low-pressure areas so that
deserts are not necessarily synonymous with the high-pressure
belts of the Tropics. The finest example is the dry-zone of the
north-west of the Indian sub-continent where the deserts of the
Sind, Thar and Baluchistan occur at the apparent focus of the
moisture-laden air from the Indian Ocean. The aridity of this area
is explained by the interference between the moist monsoon air
and the hot dry continental air which is subsiding after its long
passage over the hot, dry plateaux of Persia and Baluchistan. The
continental air, projected into and above the oceanic monsoon air
mass prevents the ascent of the moist air thus inhibiting rainfall
except for the occasional downpour when conditions of instability
temporarily obtain. The inclined plane of the Intertropical Front
reaches the land surface south of Karachi and all oceanic air is
excluded from the deserts of Baluchistan. Farther east the amount
of rain is roughly proportional to the height of the Intertropical
Front above the ground since the higher the plane of discontinuity
the higher the air can ascend and the greater the possibility of rain.
Even so, rainfall is very slight and highly irregular, producing the
arid conditions of western Rajputana and Sind. Farther south the
low rainfall of the Deccan is satisfactorily explained by the
rainshadow effect of the Western Ghats.

Of even greater interest are the paradoxical coastal deserts
which, lying adjacent to the moisture reservoir of the oceans, are
among the most extremely arid areas of the world with very low
and irregular rainfall which even extends beyond the mainland to
offshore islands such as the guano islands off Peru and South-
west Africa. These coasts and that of north-west Africa all have
extensive sections where the rainfall is less than 51 mm (2 in) per
annum with an abrupt change to heavier precipitation at the
equatorial margins. Each of these coasts is fringed by currents
moving from higher to lower latitudes carrying cool surface water
towards the equator with corresponding negative temperature
anomalies. That the cool current is a significant cause of the
aridity is shown by the higher rainfall—up to 229 mm (9 in)—of

the coast of Western Australia where the true cool current is lacking. Sporadic upwelling of cold water from the ocean depths reinforces the cool Benguela and Humboldt currents and takes the form of interlocking tongues of warm and cool water. The precise cause of the upwelling is as yet unknown though it has been suggested that its magnitude depends on the relationship between the orientation of the coastline and the current. Some theories invoke the fanning out of the surface water towards the open sea and its replacement by colder sub-surface water; others draw upon the action of offshore winds to drag the surface water seawards. Offshore winds are, however, so sporadic that they cannot be used to explain what is essentially a continuous phenomenon.

The eastern termini of the oceanic sub-tropical anticyclones show close coincidence with the western coastal deserts. These high-pressure cells are zones of subsiding air with the maximum subsidence at the eastern margin where occur the arid coastlands of Lower California, the Atacama, the Namib and the Rio de Oro. The circulation of air round the anticyclones leads to a flow of air parallel to the coast which, in turn, produces the ocean trade winds and the great swirls of the ocean currents. Subsiding air over the cool sea produces marked atmospheric stability and a strong temperature inversion, conditions which are not favourable to precipitation. Off the Namib desert in South-west Africa radio-sonde observations reveal an inversion at about 762 m (2,500 ft); a similar inversion at roughly the same altitude is found in the summer in northern California. The altitude and extent of the inversion varies locally and seasonally but it is most pronounced where the greatest contrasts between land and sea surface temperatures occur. The coldest water, frequently upwellings, appears to coincide with the greatest aridity on the land. When the inversion is close to sea-level fogs are frequent, changing to a layer of stratus cloud as the plane of the inversion rises inland.

As a result of their low latitudes these coastal deserts receive almost continuous insolation and local low pressures are created which are filled by diurnal onshore breezes. Such winds quickly die down after sunset when the ground cools rapidly by radiation. They rarely blow directly onshore since they are deflected by the Coriolis force and also come under the influence of the high-pressure cells. The deflected onshore breezes which approach the coast of the Namib from a south-westerly direction might almost be regarded as embryonic Trades or at least as local variations of the normal atmospheric circulation. Cool winds off a cool sea are the prevailing winds of the coastal deserts; as they approach the warmer land their moisture-holding capacity increases and there is but slight chance of their yielding precipitation. Inland the intense heat quickly evaporates the water particles and it is only when mountains back the coast, as in Peru, that the winds can rise sufficiently for their moisture to be condensed. Inland from the Atacama desert the zone of annual rains lies at 1,524 m (5,000 ft) on the flanks of the Andes.

Precipitation occurs when increases in temperature of the coastal waters weaken the inversion as when the divergence of the cold current from the coast is produced by a headland. At Cape Guayaquil in Peru a sharp increase in rainfall occurs, perhaps because stagnant warmer water is left by the swinging of the current away from the coastline. Similar examples occur at the bulge of north-west Africa, south of Port Etienne, and near Mossamedes in South-west Africa. Such coastal projections may also lessen the upwelling of cold water by interfering with the movement of surface waters. Irregular variations in the temperatures of the local sea surface may also explain the torrential episodic rainfall of the coastal deserts. Off Peru, for instance, the cold Humboldt current may be temporarily replaced at very infrequent intervals by *El Niño*, a warm current from the north, which produces unstable atmospheric conditions through the disappearance of the inversion. It was perhaps such a reversal of current that produced, in 1925, 394 mm (15·5 in) of rain in the lower Chicama valley in Peru where the average is only 4 mm (0.15 in) per annum. The town of Chicama was practically destroyed by floods sweeping down from the upper valley where 1,397 mm (55 in) instead of the normal 597 mm (23·5 in) of rain was received. Irrigation channels were broken and the fields covered with mud and gravel but there was some compensation since the additional moisture in the soils gave an increased yield of sugar-cane. In South-west Africa the Namib is most likely to receive rain when a temporary absence of the inversion coincides with the presence of moisture-laden air from the Indian Ocean.

Many factors may thus lead to aridity of large areas of the earth's surface. Latitude and longitude cannot, by themselves, explain the location and size of the components of the arid world; neither can continentality nor the distribution of high- and low-pressure cells, nor the upwelling of cold ocean waters. To these

must be added factors originating far back in geological time—
the formation of the continents and the ocean basins. On these
must be superimposed the more recent cycles of orogenic activity
which created geosynclines and uplifted the much modified sedimentary structures. The areas of high relief created by isostatic
uplift have been worn down and later uplifted into platforms and
plateaux by differential earth movements to create an architecture
of isolated topographic basins in the rainshadow of the adjacent
hill masses. When the western Cordillera of North America were
uplifted at the end of the Cretaceous period they were dislocated
into more humid fault blocks margining drier intervening rift
valleys and fault-angle depressions. One of the youngest arid
areas of the world is also tectonically the most recent rift valley;
Death Valley in California dates only from the Pleistocene and its
landscapes reflect its recent origin. It is perhaps significant that
almost all the world's arid lands are dominated by adjacent high
relief or have high mountains in their midst. This fact directs
attention to tectonics, erosion and sedimentation with the con-
comitant lightening and overloading of the crust compensated by
uplift and depression. Since changes in altitude are linked with the
amount of precipitation received, the importance of tectonic
movements cannot be overestimated. Conversely the zones of
sedimentation have, for many reasons, proved attractive to the
inhabitants of the arid lands since there accumulate the deposits
on which crops can be grown and on which there may be sufficient
vegetation for animals to graze. These too are the zones where the
rivers drain into interior basins providing water for irrigation or
for stock. To them man has been drawn since the dawn of prehistory and repeatedly failed to conserve the moisture which was
available. In more extreme cases he has, in the process of estab-
lishing himself, even extended the area of the arid lands and thus
himself become responsible for causing aridity. At this stage it
becomes essential to look more closely at the characteristics of the
climates of the arid lands, to understand the natural processes
behind the environment which man is now attempting to develop.

THE CLIMATES OF THE ARID LANDS

Significant variations in the character of the world's dry lands have already been established using criteria based on rainfall, vegetation and soils. Desert and steppe, arid and semi-arid, betray the combined effects of rainfall and evaporation yet a further basic distinction remains, namely between the hot and temperate deserts with their associated steppe lands. Since summer temperatures are high in both types, the terms hot and temperate apply to the thermal characteristics of the winter half-year and it is to be expected that latitude will be mainly responsible for this variation.

The hot deserts

With an annual average temperature of over 18°C (64·4°F), the hot deserts typified by the Sahara, Libyan, Arabian and Great Australian deserts fall into the BWh category of Köppen's classification. On the Equator and poleward sides they are transitional to the hot steppe lands (BSh) where the rainfall is not only more plentiful but has also a prominent seasonal maximum. Occasionally the hot desert may grade directly into temperate desert as in the arid south-west of the USA and in Central Asia.

Lying roughly within latitudes 20° and 25° north and south of the Equator but with marginal extensions 5° or more beyond these limits, the hot deserts occur principally where the subsiding and diverging air masses of the high-pressure cells provide atmospheric conditions unfavourable to precipitation. Such areas are characterised by very high summer and high winter temperatures, large diurnal temperature ranges but only moderate annual temperature variations. Evaporation is high and relative humidity low. There is a high sunshine and low cloud amount; a hot sun beating down from a clear blue sky is the normal weather and climate in one. Wind speeds are moderate and usually regular in force and direction. Variations in any of these characteristics provide such diversity as is present in the hot arid zones.

The highly variable rainfall makes nonsense of mean annual rainfall statistics which, as the figures from the selected stations in the table below indicate, are always low in amount.

Alice Springs (Australia)	252 mm (9·93 in)
Charlotte Waters (Australia)	147 mm (5·08 in)
Jacobabad (W. Pakistan)	102 mm (4·0 in)
Aden	48 mm (1·9 in)
Cairo	30 mm (1·2 in)
Tamanrasset (Sahara)	41 mm (1·6 in)
Ghardaia (Sahara)	61 mm (2·4 in)
Karibib (South-west Africa)	183 mm (7·2 in)
Yuma (Arizona)	84 mm (3·3 in)

That the mean annual rainfall figures are unreliable is shown by the data indicated earlier for Yuma which in 1899 experienced only 25 mm (1 in) of rain, but more than 280 mm (11 in) in 1905. At Tamanrasset 160 mm (6·3 in) fell in one year and, in another, only 6·4 mm (0·25 in). Although rain does not necessarily fall every year the areal discontinuity of rainfall makes it dangerous to be dogmatic about complete drought for any portion of the earth's surface since a rainfall belt may be only a few square miles in extent and not recorded a few miles away. On the other hand, the extension of regular air-line services over the Sahara has revealed narrow bands of rainfall extending for hundreds of miles. Such a rain belt extended, in 1942, from Dakar in West Africa to southern Morocco. It is hoped that photography from satellites may reveal the exact distribution of these rainfall zones. In the absence of long-term meteorological data, it is difficult to decide whether prolonged periods of drought occur; the use of oral tradition has proved unreliable since nomadic tribes tend to ignore ineffective rain which fails to penetrate the soil, produce grass or replenish wells. Areas for which no rain was remembered over a six-year period showed unmistakable evidence of flow of water in the dry river beds during the same period. According to Capot-Rey, there are no districts in the western Sahara which fail to receive important rainfall in a ten-year period or go for more than six years without receiving more than 5 mm (0·2 in).

Agriculture without irrigation is clearly impossible under these conditions and even nomadism based on stock grazing is highly precarious. The Touareg of the Ahaggar highland in the Sahara have sometimes been forced to migrate as far as the Sudan in the search for grazing: this indicates that the slight increase in rainfall of the higher areas of the Sahara—about 102 mm (4 in) yearly average—is not necessarily accompanied by marked increase in the regularity of the precipitation.

Rain in the hot deserts usually falls in convectional showers of short duration and limited areal extent on a small number of rainy days. In the extremely arid areas they have no seasonal regularity. Gentle showers are not unknown, but since they scarcely moisten the soil or provoke run-off before being evaporated, they have less effect on fluvial erosion than the torrential showers although their contribution to the desert weathering processes and to the growth of vegetation may be all-important. It is, however, the torrential downpour which is remembered by oasis dweller, pastoralist and traveller. Such rains create havoc by destroying mud walls and houses (in 1922 twenty-two women were buried by the collapse of a wall at Tamanrassset) and by sending swirling torrents down the wadis of the Sahara or the *arroyos* of the North American deserts. In 1919, after a prolonged period of drought, local floods occurred in Cairo when 43 mm (1·7 in) of rain fell in one day and trams were buried in mud up to the window-sills. This is the rain which replenishes ground-water supplies, introduces fluvial processes to the desert landscapes and, by moistening the soil, gives an opportunity for the short-lived annuals to grow. Tamanrasset, with 40 mm (1·6 in) mean annual rainfall, has received 44 mm (1·7 in) in three hours of which nearly three-quarters fell in forty minutes. In the Tibesti highlands of the Sahara the military post of Aozou received 370 mm (15 in) of rain in three days in May 1934 which produced great floods in the wadis. At Doorbaji in the Thar desert where the annual rainfall is only 127 mm (5 in), 864 mm (34 in) has fallen in two days. Damascus—mean annual 234 mm (9·2 in)—received, in February 1945, 76 mm (3 in) in one morning, roughly half the precipitation for the whole of 1945.

Such concentration of the rainfall into a small number of rainy days is highly characteristic of the arid zone. Charlotte Waters in

Central Australia—annual rainfall 130 mm (5·1 in)—receives its rain on only twenty-five days per annum, giving an average of 5 mm (0·2 in) per rainy day. At Suez and Cairo rain falls on only ten or eleven days per annum to give average daily falls of 2·5 mm (0·1 in). Apart from the areas near the desert margins, rain falls at no particular season and rain in, say, May in one year may be followed by rain in December the following year.

Within a hot desert area such as the Sahara altitude has some effect on rainfall amounts. The isohyets bend equatorwards to coincide with the massifs of Tibesti and Air, while, to the north-west, the Ahaggar, with its cuestas of the Ennedir and the Tassili des Ajjar, stand out as pluvial islands, although amounts are still only of the order of 102 mm (4 in) per annum. In contrast, the great depressions of the Sahara are very dry. The Erg of Mourzouk, east of the Ahaggar, sheltered on three sides by high ground, has only about 10 mm (0·4 in) mean annual rainfall while the Kufra basin in the Libyan desert is probably the driest district in the whole of the North African desert tract.

Aridity in the hot deserts depends not only upon sparse and irregular rainfall but on high temperatures and high evaporation rates all the year round. In summer the insolation is intense with the sun above the Tropic. It becomes lower on the equatorial margins of the Sahara and the Great Australian desert where there is cloud associated with the summer rainfall regimes of the hot steppe and savanna. The wisps of high cirrus over the Sahara, the 'counter-trades' at over 1,829 m (6,000 ft), offer little hindrance to the fierce rays of the sun which raise ground temperatures to extraordinary values. The Sonora desert in the south-west USA and Mexico experiences about 90 per cent of the possible sunshine in summer and Yuma in Arizona has only one-twentieth cloud cover in July. For Yuma the total insolation in the course of the year is 3,900 hours or 89 per cent of the maximum possible. In comparison the humid tropical forest zone of, say, the Congo may receive only 1,800 hours of insolation per annum, or less than half that received at Helwan in Egypt.

Everywhere the soil temperatures are higher than air temperatures. Sand, rock and metal may reach mid-afternoon temperatures of over 82°C (180°F) and can be felt through stout boots and thick-soled sandals; a hand placed inadvertently on a car bonnet is quickly withdrawn; water, unless kept cool in earthenware pots or canvas containers to permit cooling by evaporation, does little to quench the thirst; although refrigerators and air-conditioning have now made life tolerable in oil prospectors' trailers, the tractor vehicles require special cooling arrangements to conserve water or other coolant. Unshod horses suffer severely but the thick foot-pads of the camel offer greater protection. It is obvious that desert plants must be provided with heat resistance mechanisms if they are to survive such high soil temperatures. One also wonders how many soldiers in the campaigns on the North-West Frontier, in the Foreign Legion and in the Eighth Army in World War II were killed because they could not lie flat on the burning rock or sand.

Air heated in contact with the ground sends the shade temperatures soaring to record heights. At Insalah in the Sahara temperatures of 54°C (129°F) have been recorded under standard meteorological conditions but this figure is exceeded by Death Valley in California, 84 m (276 ft) below sea-level, with 57°C (134°F), and by Azizia, 40 km (25 miles) south of Tripoli in North Africa, which holds the unenviable record of 58°C (136·4°F). Even higher shade temperatures are probably attained in many areas, but the scarcity of recording stations at the moment makes for uncertainty on this point. Over wide areas temperatures of 38–41°C (100–105°F) are easily reached and maintained for several days at a time. In the western Australian desert the thermometer recorded 38°C (100°F) on sixty-four consecutive days and 32°C (90°F) on 150 consecutive days. Summer daily temperatures at Alice Springs attain 38°C (100°F) almost continuously except during periods of rain or when a cool breeze from the south reaches the area. Similar examples are available from stations in the Sahara and Arizona. Under such conditions life and work become difficult; the eyes, peering half-closed against the glare reflected from sand and rock and shielded by peaked cap or *burnous*, become accustomed to the mirages of blue lakes suspended shimmering and undulatory.

Under the influence of intense radiation into a cloudless star-filled sky, the summer night offers some respite when temperatures fall but still only to levels higher than those of a West European summer afternoon. Night air temperatures remain above 21°C (70°F) at Phoenix, Arizona, and reach higher levels at Saharan stations. At Insalah the average July daily minima are over 30°C (86°F). In the summer the large diurnal temperature range is thus the result of high maxima rather than of low minima. Using daytime shade temperatures the daily range is about 17–22°C (30–40°F), although exceptional values have been recorded. For Death

Valley in August 1891 the mean diurnal range was 35°C (64°F) and the maximum 41°C (74°F).

Ground surface temperatures may have a range of over 39°C (70°F). In the Erg of Mourzouk the sand surface temperature on 9 April 1944 was 10°C (50°F) at midnight and 45°C (113°F) at mid-day. At a depth of 3·5 cm (12 in) below the surface, however, the temperature remained fairly constant at about 25°C (77°F) with slightly higher values in the afternoon. Accordingly the desert is populated with burrowing animals and man has discovered the advantages of living in a cave or in a converted rock-tomb, as at Cyrene in Cyrenaica. The thick walls of mud houses, or the bee-hive huts of the Syrian high plains between Aleppo and Orontes, aim also at natural air-conditioning with free circulation of the air. Air trapped in houses, however, rarely falls below 30°C (86°F) at night in the Saharan oases and it is far more comfortable to sleep in the open where the air in contact with the sand is cooler than in houses or palm-groves.

Arab clothing is designed not only to give shelter from the heat in summer but also to protect against the chill of the winter nights. Biskra in the north of the western Sahara, just south of the Atlas mountains, has average January temperatures of 11°C (52°F) and minimum temperatures for the same month of 6°C (43°F). At Agades, on the southern side of the Air massif, the January mean is 20°C (68°F). Throughout the hot deserts winter temperatures show more variation than in summer when the overhead sun produces much greater thermal uniformity. In the Saharan, Arabian, Australian and American deserts the winter isotherms run roughly from east to west, reflecting latitudinal controls.

Daytime temperatures are lower than in summer but the night temperatures are more interesting since they drop to low levels and sometimes even below freezing point. At Yuma, with a January mean of 12°C (54°F), the maxima and minima recorded are 27°C (81°F) and −6°C (22°F). An American desert resort such as Las Vegas can have the best of both worlds by air-conditioning for the hot summer day and cold winter night but the Old World nomad is not so fortunate. On the lowlands the night temperatures in the deserts of the Middle East do not always descend to freezing point but huge fires are lit when fuel is available and blankets and a tent are a necessity. T. E. Lawrence records how sugar for the first cup of coffee in the chill of the dawn was not considered effete for the men of the hardy Bedouin. Over the greater part of the Arabian desert, except in the south and south-east, frost and snow may be experienced while a definite frost season exists at Alice Springs in Central Australia between roughly the middle of May and the end of August. In the corridor between the Erg Occidental and the Erg Oriental in the Algerian Sahara at El Golea, frosts cause damage to citrus fruits and restrict the growth of winter-active plants. When temperatures range from those of an English summer's day to those of an English winter's night the diurnal air temperature range is obviously high but it is exceeded by the diurnal variation in soil temperatures in the uppermost layers.

Altitude lowers temperatures still further. The palm tree is absent from the Ahaggar in the central Sahara owing to the regularity of winter frosts. At Tamanrasset, over 1,219 m (4,000 ft) above sea-level, the January mean is 12°C (53°F) and the absolute minimum −7°C (20°F). At a greater altitude in the Colorado basin than Yuma, Fort Grant at 1,498 m (4,916 ft) above sea-level has recorded −12°C (10°F) and at Santa Fe in New Mexico, at over 2,133 m (7,000 ft), the temperature dropped to −25°C (−13°F). That increase in altitude affects the incidence of frost is also indicated in central Australia where Alice Springs, 584 m (1,916 ft), has approximately 102 days with frost, while Charlotte Springs, 213 m (700 ft) above sea-level, has an average frost season of sixty-four days.

It has been said that the mummies of Egypt were preserved almost in spite of the efforts of the Egyptian embalmers, by the dryness of the sand and the atmosphere, and similar mummification is recorded by archaeologists in the arid areas of coastal Peru. The relative humidity of the air is usually low in deserts and values of as little as 2 per cent have been recorded; on 5 October 1951, when a south-east wind was blowing, the relative humidity at Jalo in the Libyan desert was only 9·5 per cent and the shade-temperature 41·1°C (106°F). Under such conditions the lips are parched, the skin dry and thirst almost insatiable. Water with the high content of 3·88 parts per thousand of dissolved salts is as welcome as water from the crystalline rocks of upland Britain although the effects on the digestive tract are somewhat disastrous.

Relative humidities are, however, generally higher than the values already indicated. For Alice Springs they range from about 36 per cent in winter to 25 per cent in summer; for Yuma the values are 47 per cent and 34 per cent respectively, but are not strictly comparable since the readings do not refer to the same time of the day. Variations in the relative humidity result principally from changes in wind direction. Within a short period in

October 1951 the relative humidity at Jalo jumped from 9·5 per cent to 85 per cent as the south-easterly wind changed round to blow from the north. At Cairo, the *Khamsin*, a desiccating southerly wind associated with the movement in winter of depressions along the Mediterranean, lowers the relative humidity to under 25 per cent but it rises to 80–85 per cent when the winds blow off the Mediterranean. As long as the relative humidity is not too low, nocturnal cooling of the air near the ground below the dew-point provides precious extra moisture which may support vegetation and mists often form in valleys and depressions. As temperatures rise during the day the dew is quickly evaporated but not before it has wetted rock surfaces and supplied the essential moisture for weathering processes. Low relative humidities require low dew-point temperatures for condensation and if the relative humidity is, say, 30 per cent as at Alice Springs, the temperature must fall almost to freezing point before dew is deposited, a condition which is unlikely to be fulfilled in the summer months.

High temperatures and low relative humidity lead inexorably to high evaporation rates although the scarcity of recording stations with evaporimeters makes it impossible to give an accurate and comprehensive picture for the deserts as a whole. At Alice Springs the evaporation rate amounts to 2,413 mm (95 in) per annum, roughly ten times the average rainfall; stations in the northern Sahara record an evaporation rate of up to 4,064 mm (160 in) of evaporation per annum. Yuma during the summer has 1,397 mm (55 in) of evaporation and 25 mm (1 in) of precipitation.

The quantitative study of evaporation losses was given an ideal natural setting when, in 1904, the Salton Sea in the Colorado desert was formed by floodwaters from the Colorado River. By 1907 the river had again been confined to its proper channel leaving the 1,140 sq km (440 sq m) of the Salton Sea as a freshwater lake. Evaporation losses were determined by measuring the inflow, discharge and water level and showed an average of about 1,524 mm (60 in) per annum. More recently studies of lakes in the Jordan river system including Lake Tiberias and the Dead Sea have provided, as has Lake Mead in North America, valuable information for the engineers engaged in the problems of storing drinking and irrigation water in open reservoirs in the arid zone.

The hot coastal deserts
The characteristics of the hot deserts are in some ways modified along the western coasts where either cool currents or the up-

welling of cold water introduces significant variations. Temperatures are affected by the cooling influence of the sea and annual
temperature ranges are lowered considerably. The average annual
temperature at Callao in Peru is only 19°C (67°F), at Iquique in
Chile 19°C (66°F); both these values are anomalous for the
latitude. The hottest month at Callao is only 22°C (71°F) and the
coldest 17°C (62·5°F), an annual range of 5°C (8·5°F). At Walvis
Bay in South-west Africa the annual mean temperature is 17°C
(62°F) and the average annual range only 6°C (10°F). The difference between the mean maximum for the hottest month and the
mean minimum for the coldest month is only 16°C (29°F). The
daily range at such stations is only about 11°C (20°F), about half
that of stations in the interior of the Sahara, whose western
coastal fringe of the Rio de Oro shows similar modifications in
the temperature regime.

In the amount and character of the precipitation the coastal
deserts do not differ appreciably from hot deserts in inland situations. The arid zone of coastal Peru receives only about 25 mm
(1 in) average annual rainfall—Callao has 30 mm (1·18 in). In
South-west Africa, Swakopmund receives only 16 mm (0·65 in).
The rainfall is also very irregular both in seasonal incidence and
annual amounts. The case of Chicama was cited above (page 22)
where long drought periods are broken by torrential showers.
Similarly Swakopmund was drenched in 1934 by a shower of
51 mm (2 in) in one day; for the whole of that year the rainfall
amounted to 155 mm (6·13 in). Further inland on the Namib
platform about 731 m (2,400 ft) above sea-level and 80 km
(50 miles) from the coast, the annual mean rainfall is only 36 mm
(1·39 in) yearly, yet amounts vary from 76 mm (3·1 in) to 15 mm
(0·6 in). The keynote is low, highly variable rainfall.

Distinguishing the coastal from the interior deserts is the high
relative humidity and fog produced by prevalent onshore winds.
This high humidity can support meagre plant life and produces at
an altitude of 1,524 m (5,000 ft) a belt of permanent vegetation in
the Andes of Peru. The relative humidity at Walvis Bay during
the first half of a midsummer's day stands regularly at 100 per
cent. After midday, as the temperature rises and the fog and
stratus break up, the relative humidity is lowered to about 75 per
cent. On some days the variation is only 10 per cent, ranging from
a maximum of 100 per cent to a minimum of 90 per cent. At
Swakopmund fog is recorded on 150 days per annum, while off the
west coast of South America Darwin recorded in his *Voyage of*

the Beagle that he saw the Cordillera behind Lima in Peru only once during the first sixteen days because of the stratus layer which is moved inland by the sea breezes and later evaporates. The sea breezes on such coasts increase in velocity with the heating up of the land until, by noon, they blow at Force 3 on the Beaufort scale. For most of the afternoon they have a velocity of Force 4 or 5, and lift the finer particles from the drying-out sand surface to make life very unpleasant.

In South-west Africa the normal pattern of sea-breezes may, however, be broken by the *Berg* wind which is a very hot and dry offshore wind, and locally called the 'Doctor'. This carries great quantities of dust many miles out to sea and interferes with the movement of shipping. It has, however, the merit of drying out the dampness caused by the normal high humidity and raises the temperatures to values of over 32°C (90°F). The maximum temperature experienced under *Berg* wind conditions in the Namib was 46°C (115°F) at Port Nolloth. Evaporation, which is normally low, can reach high values under *Berg* wind conditions and normally increases inland as the mists rise away from the coast.

The hot steppe lands

The steppe lands offer the greatest problems in delimitation and utilisation. If most people can recognise absolute or extreme aridity, the pages of the history of land-use in both Old and New Worlds and in Australia indicate that semi-aridity is not so easily recognised since there are numerous examples of the mistakes made by the agricultural and pastoral occupants of the semi-arid lands. The steppes are essentially transition zones which combine the characteristics of the true desert climate with those of more humid regions. The boundary between the steppe and more humid climates is often placed where evaporation and precipitation are equal in amount although the equilibrium may vary from season to season and from year to year. Compared with the humid margins precipitation is lower and more variable and evaporation is higher, but the seasonal incidence of rainfall and the temperature characteristics are maintained into the drier climate zone. It is this marked seasonal incidence of rainfall which really distinguishes the semi-arid from the truly arid lands.

In Köppen's classification, the letters BSh designate the hot steppe with the suffixes 'w' and 's' to indicate the driest period in winter and summer respectively. North Africa and Australia show the best development of the hot semi-arid lands in relation to the

B

hot deserts but extensive areas are found also in West Pakistan and southern Iran and in the northern and eastern margins of the Namib and Kalahari in southern Africa. The hot steppes of North America are found principally in Lower California and Mexico. The drier north-west and the rainshadow area of the Deccan of the Indian sub-continent should be included in the hot steppe lands. Sections of the great valley of California would also come into this category.

In North Africa the margins of the Sahara offer examples of the transition to more humid climates with rainfall at contrasting seasons of the year. Lying on the northern side of the Sahara is a zone of steppe lands transitional to the Mediterranean climate proper. This belt receives its rains from the cyclones developed along the surface of contact between polar and tropical air which is aligned approximately along the Mediterranean Sea in winter. It is also possible for cyclonic rain to fall at this season along the southern steppe margin of the Sahara transitional to the savanna lands.

The winter precipitation of these steppes north of the Sahara amounts to as much as 508 mm (20 in) and some may fall as snow at higher elevations. Rainfall variability is high but the average rainfall figures are more meaningful than in the deserts. Tozeur in Tunisia has 89 mm (3·5 in) while Sfax on the coast has almost twice this amount. Although low in quantity this rainfall is effective for vegetation since it falls in the winter when temperatures are lower and evaporation less. Temperature ranges are less than in the Sahara with average maxima about 26·7°C (80°F) on the coast and average minima about 12·8°C (55°F). Frequent changes in wind direction during the passage of depressions in winter cause rapid variations in humidity, notably when the southerly winds prevail. These have different names in different areas but the *Khamsin* of Egypt is one of the best known.

The hot steppe on the southern margin of the Sahara in the Sudan and West Africa, the Sahel, is made more arid by the incidence of rainfall during the summer when temperatures and evaporation rates are high although a little lower than those of the Sahara itself. Extremely high temperatures and very low relative humidities may be exported from the Sahara by the *Harmattan* which desiccates still further the already dry environment. Within the Sahara, the Harmattan, blowing from the north-east, can raise violent dust-storms which obscure visibility and penetrate through clothing into eyes, ears, nose and throat. To the southern steppe zone it imports clouds of dust and extends the mobile sand-dunes towards the south-west. Annual temperature ranges are lower than for the semi-arid belt to the north. At Hillet Doleib in the Sudan the range is only 5°C (26°–31°C) or 9°F (79°–88°F). Rain falls as the moist equatorial air masses move north during a few months of the summer and the displaced equatorial trough briefly establishes itself through altitude; the occurrence of rain in the hot season reduces its effectiveness so that Hillet Doleib, with 762 mm (30 in) falling in months when the average temperature is over 26°C (80°F), has the same measure of aridity as Tozeur in Tunisia where the rainfall is only one-tenth of that amount.

The temperate deserts

The challenge of aridity extends beyond the area dominated by the tropical high-pressure cells to the interiors of the continents in middle latitudes. Vast areas of interior Asia and North America are either temperate deserts or steppes. Since summer temperatures may be as high as in the hot deserts, the distinguishing characteristic must be the intense winter cold which brings in its train major changes in vegetation and utilisation. The temperature differences between summer and winter in these temperate deserts offer some of the largest annual ranges on the earth's surface and the winter cold may be intensified by cold winds or by cold air draining down the slopes of the mountains which often surround the more arid basins. Temperatures in winter may fall as low as −1°C (30°F) below zero producing frost in the soil and the possibility of snowfall. Kazalinsk (in Kazakhstan SSR) has 70 days with snow and 183 days with frost. At Tashkent where the average January temperature is −1·3°C (29·7°F) the temperature may fall to −30°C (−22°F) during cold waves; snow may fall on thirty-seven days in the year with a frost incidence of 152 days. Although such conditions are characteristic of much of Central Asia, winters in areas in the same latitude such as the Caucasus are much milder with little snow and few frosts. Similar variations are found in North America where parts of Arizona and Colorado have January minima above freezing and consequently come within the category of hot desert although the higher plateaux basins and valleys of Arizona, Colorado and New Mexico may often register −17°C (30°F of frost).

These temperate deserts are characterised by high annual and diurnal temperature ranges, dry air and high insolation through an atmosphere little impeded by cloud. Orographic effects are often very marked with a coincidence of arid and semi-arid conditions with basin and range relief. In Central Asia the principal desert basins are open to the north from which flows cold air in winter which lowers the temperatures appreciably, giving much lower January minima than in the North American temperate deserts. Average monthly temperatures for the winter half-year are below freezing. The Aral Sea is frozen for four to five months each year and in many years the ice floes can persist until mid-May. The lower section of the Syr Darya in southern Kazakhstan is frozen (here the January temperature of −12°C (10·4°F) is lower than in the Gulf of Finland) yet while there is still snow on the lower Syr Darya the peach and almond trees are in bloom in Tashkent. Farther to the east Lake Balkhash occasionally freezes over completely. Cold winds from the north and north-east frequently blow the deep snow into drifts producing a completely hostile environment. Daytime maxima may rise on calm sunny days to about 4°C (40°F) but temperatures such as this are more common in the more southern arid tracts where even the coldest months of January and February have temperatures above the freezing level, except when cold winds from the north intervene bringing a little snow which lies for only a short time.

Summer in contrast is exceedingly hot and dry. At Bairam-Ali in Turkmenistan ten consecutive summers (July to September) have occurred without a single drop of rain. The great heat and low precipitation, low cloud amount and relative humidity combine to produce a desiccated, dusty environment. Whirlwinds, produced by turbulence due to the unstable condition of the lower layers of the atmosphere in contact with the heated ground, reduce visibility and lower the transparency of the atmosphere. On some days in the southern desert zone shade temperatures approach 50°C (122°F) and ground temperatures are correspondingly higher. In southern Turan the summer daily maximum in the shade exceeds 50°C (122°F) while at Kepelik, on 20 July 1915, the sand surface temperature reached 96°C (174°F), and the daily range of 78°C (140°F) at the sand surface is not uncommon. Conditions are analogous in fact to the hot deserts of the tropics, and demonstrated by the burrowing habits of desert animals during the day with emergence for feeding during the cooler night when temperatures are, on average, 25°C (45°F) lower and may be as much as 50°C (90°F) less than in the daytime.

Dry dusty winds may blow with considerable strength, increasing in velocity in the afternoon. The *Afganets*, a significant local wind in the upper Amur Darya, blew on 20 to 22 August

1930 at Termes and raised so much dust that the sun could safely be observed with the eyes unprotected by dark glasses.

In association with the low cloud amount in summer, when high cirrus does little to obscure the sun's rays (there are more hours of sunshine in Central Asia in summer and autumn than in Egypt), and the low values of relative humidity which averages only 30 per cent in the south and may drop to 5 per cent, evaporation reaches high intensities with a July maximum. Tashkent has an evaporation rate which exceeds three times that of the precipitation on average but in individual years the disparity may be even greater. Turtkul where evaporation exceeds precipitation by 36 times in average years has on occasions 270 times more evaporation than precipitation which over the greater part of Central Asia is less than 254 mm (10 in) per annum. Although the actual amount of water evaporated from the dry desert surfaces is obviously low, evaporation from lakes and rivers, where water is available, is high. The Syr Darya at Kazalinsk loses by annual evaporation 1,463 mm (57·6 in) of which nearly three-quarters is lost between April and September. Since the conditions in summer approach so closely those of the hot deserts it is necessary only to stress their similarity both in Asian and North American situations, although regional relief as in the basin of Fergana may cause significant local differences.

The temperate steppe lands

These areas in North America and Asia form part of the transitional range between the deserts proper and the continental grassland areas. They have greater precipitation than the deserts but the rainfall is still unreliable. These have been the regions of great economic catastrophe where the reality of semi-arid conditions was not always perceived or where the environment was nevertheless treated as if it lay in the humid zone. Within this category fall the Great Plains of the USA between the normally dry lands to the west and the normally wet lands to the east. They proved to be a zone of climatic fluctuation with years of drought when the dry margin advanced towards the east. In Central Asia the temperate steppe grades into the semi-desert tract extending from the Emba River, which drains into the Caspian, to Lake Zaysan in the east, and continues through Mongolia along the northern margin of the temperate desert of the Gobi into the basin of the Hwang-Ho and north-west China.

As in the temperate deserts sharp annual and daily fluctuations of temperature occur. A short spring separates the cold winters from the hot summers when the small and unreliable precipitation —less than 304 mm (12 in)—mainly occurs. Winter precipitation usually falls as snow but both in Central Asia and in the semi-arid lands of the American west it forms only a thin cover which is easily melted in warmer spells. When followed by freezing conditions a layer of ice may form which prevents animals from reaching the low-growing semi-arid vegetation. The remaining snow, other than that compacted into snow-drifts, melts quickly with the rapid increase of temperature in the spring and, augmented by the spring rains, either runs off rapidly over still frozen ground to the rivers or is evaporated leaving but a small residue to soak into the ground. The delayed melting of the snow-drifts is thus a valuable source of water for plants in the spring.

Summer temperatures do not normally rise as high as in the deserts but shade temperatures of 40°C (104°F) are likely to be reached and their effect is accentuated by the dry winds which ensure high evaporation rates. Summer rains falling on ground thinly covered with vegetation wash down the gullies and little goes to store as ground water.

Local and microclimates

Large tracts of the earth's surface are arid lands with relief and land forms ranging from high plateaux and mountain basins to alluvial plains. It is thus necessary to consider the possible variations in local and microclimates. Minor variations in relief, altitude, aspect and humidity have far-reaching consequences for animals, vegetation and human beings. The zone of interest includes not only the layer of air near the ground but also that of the air in surface materials. Since the system of relationships between radiation, temperature, humidity, precipitation and wind may be disturbed by the introduction of factors which interfere with any one of them, it is clear that completely new microclimates may be introduced by irrigation, windbreaks, cropping and dry-farming practices. Such new microclimates may, for instance, provide conditions favourable for the breeding of pests which under natural conditions were restricted to other parts of the environment.

Evapo-transpiration values can vary widely within short distances. Measurements in the Egyptian desert showed that in an area of artificial shade the value was only half that on the exposed plateau; in a wadi nearby it was three-quarters while the lowest figure was obtained within a cave a few metres from the entrance. The presence of mulch at the soil surface, consisting of a layer which differs in its characteristics from that of the underlying soil, can reduce evaporation, and is used as the basis of dry farming. A mulch of stone or gravel helps to reduce transpiration loss by reducing the growth of weeds. Natural mulch produced by constant tillage has a beneficial effect by reducing the maximum temperature in the root zone of crops and is also believed to conserve moisture in the underlying layers. Wind-breaks such as the sides of gullies, sand-dunes or vegetation can increase water conservation both to windward and in their lee and influence soil and air temperatures markedly. A change in the reflectivity of the surface through vegetation, gravel, sand or rock exposures alters the heat balance and the microclimate of the soil and the air near the ground. Such a variation in the surface materials may also have profound consequences on the incidence of dew which, although small in total—10 to 20 mm (0·4 to 0·8 in) per night in the arid zones—if spread over a large area, may have important consequences for animals and vegetation when localised in areas of low thermal conductivity. Here the night temperatures would be lower than the surrounding areas provided that the air was still or the winds were very weak and the amount of dew precipitated would be much higher than the uniform rate quoted above. Leaf-litter and patches of stones cause dew concentrations and may have been so used in the past for cultivation in the arid zone. Although comparatively little is known about dew formation there is a body of opinion which prefers its derivation from evapotranspiration rather than from condensation of air-mass humidity, in other words the conservation of moisture already locally available rather than the introduction of a new supply.

It is clear that an oasis with palm trees or an irrigated field creates a local climate within the general climatic framework. An oasis is, in effect, an ecological island with conditions quite different from the desert on its margins. The burrowing animal also creates its own microclimate insulated from the climate near the ground under the harsh conditions of which it could not survive in the daytime. An irrigation canal crossing a desert can act as an ecological routeway for the migration of insects not adapted to the desert environment. In Central Asia the irrigation canal provides a cool moist local climate and is favoured for the sites of shops selling cups of tea. They are situated on the side of, or bridged across, the irrigation canals to take advantage of the cooler air whose higher humidity may help to preserve the flavour of the leaf. The ploughing of virgin semi-arid lands alters the structure and microclimate of the soil and by the change in the spacing of vegetation from a close to an open cover upsets the whole of the ecological balance. Forest clearance, overgrazing and fallowing lead to changes in the soil structure and its climate as well as climate of the air layer in contact with it. The development and utilisation of the arid zone have therefore involved a series of consequential ecological changes which transformed the 'natural' climate to a greater extent than in the more humid zones and may have even created aridity by destroying the ecological balance.

Climatic change

Present-day climates of the arid zones of the Old and New Worlds are not sufficient to explain many of the physical and human phenomena. Relief, vegetation, soils and the history of land utilisation indicate that in earlier periods of geological time, and since man's appearance on the earth, there have been profound changes in the amount and reliability of the precipitation received and in the rates of evaporation. Rainfall variability in the arid zones has already been stressed and with the increasing amount of evidence now available it is clear that there have been, and will be, long- and short-term climatic fluctuations. It is also clear that man himself has been responsible for changes in the extent of the arid area and its physical characteristics through the introduction of grazing animals, burning and tillage. Earlier controversies concerning the part played by climatic change and human intervention are gradually being resolved and it is now possible to summarise the main conclusions reached by investigators in the Old and New Worlds.

The evidence for climatic change is based on the findings of palaeoclimatologists, geologists, botanists, archaeologists, soil scientists and geographers. They have studied the rise and fall of lake levels, the fluctuation of the ground-water table, the cycles of sedimentation and erosion, the occurrence of relict landforms and paleosoils and dated them by both relative and absolute methods using, among others, the study of varves (the annual layers of sediments deposited in lakes, especially on the margins of glaciers in the Quaternary Ice Age) and tree rings, Carbon-14 and the artefacts of prehistoric man. The occurrence of wetter periods in the arid zone is well established but it now seems probable that there has been little significant change in the climate of the arid lands during the last 2,000 years which focusses attention on man's activities for changes in the environment in the historical period. From considerations of the basic atmospheric circulation patterns it also seems probable that the sub-tropical arid zone was never completely eliminated in prehistoric and geological time but was simply reduced in extent with some latitudinal shift.

The changes in climate which are most important in a study of the utilisation of the arid zone are those which have occurred during the last million years since the end of the Tertiary era. In the Pleistocene period a great part of the Northern Hemisphere experienced a major change from the world climates which had prevailed during the greater part of geological time. A succession of well-established glacial periods, the Gunz, Mindel, Riss and Würm with their North American equivalents of the Nebraskan, Kansan, Illinoisan and Wisconsin, led to the formation of ice caps over much of northern Eurasia and North America. Ice sheets spread across northern Europe from the mountains of Scandinavia while from secondary centres in the Alps, Pyrenees and Carpathians ice spread for more limited distances to the north and south. In the areas not covered by ice severe sub-Arctic climates left ample evidence of their existence in the form of fossil frozen ground phenomena and valleys filled with sediments. In North America ice spread south from the Laurentian Shield to approximately the latitude of New York on the east coast and secondary ice caps were established on the western mountain system of the Cordillera. Two glaciations are recorded in the East African Highlands of Abyssinia and Mount Kenya. Not only was the Pleistocene a period when a climatic refrigeration occurred; it also saw the first appearance of man for whom a considerable part of the present world's living space was then not available because of the extremely hostile environment. It is also well established that the periods between the glaciations, or the interglacials, enjoyed a climate which was warmer and moister than that of northern Europe and North America at the present day. Prehistoric man moved north or south, following the animals on which he depended for subsistence.

Nearly twenty years before the end of the nineteenth century, not long after the universal scientific acceptance of the reality of glaciation in the northern continents, it was postulated that the glacial epochs in the north bore a direct relationship to periods of moist climate in areas in lower latitudes which are now arid. During the Wisconsin glaciation in North America a lake of

1,165 sq km (450 sq miles) in extent and 46 m (150 ft) deep occupied
the closed Estancia valley in New Mexico in an area where the
present average precipitation of 356 mm (14 in) is quite insufficient, in view of the high rates of evaporation, to maintain
such a body of water. At least 508 to 610 mm (20 to 24 in) of precipitation was required to create a lake of this volume. G. K.
Gilbert's classic work on the former shorelines of Lake Bonne-
ville and numerous other examples testify to the greater precipitation in these arid parts of the USA where present-day rainfalls
are less than 102 mm (4 in) per annum. In Eurasia during a glacial
period of the Pleistocene the Caspian Sea overflowed into the
Black Sea, although, in this case, it is believed that the water-level
rose as a result of lower evaporation rates derived from lower
summer temperatures rather than from increased precipitation.
In the southern margin of the Sahara where lowering of temperatures during the glacial periods was negligible, increased precipitation resulted in fresh-water lakes evidenced by lacustrine
deposits and Lake Chad attained vast extent and depth. Lakes
also existed in the Sahara proper between the highlands of Tibesti
and Ennedi while in Australia there is evidence from former lake
deposits of a much enlarged Lake Eyre which testifies to more
favourable moisture conditions than exist in that area at the
present day. The Dead Sea, situated during the glacial periods in
a zone with lower summer temperatures and increased length of
rainy season, was nearly 228 m (750 ft) deeper than its much
reduced and saltier successor. The importance of the pluvial
periods, or pluvials, to present-day occupancy of the arid zone
lies, however, in the water which filled the water-bearing rocks or
aquifers and provides, in dwindling quantities, the ground-water
on which many cultivators and pastoralists depend.

The pluvials, defined as a long-term humid fluctuation extending
over a considerable area, were like the glacial periods, separated
by 'interpluvials' with less precipitation so that the Pleistocene
was by no means one long moist period in lower latitudes.
Such interpluvials have been correlated with the interglacial
periods. The 'interpluvial' climates were not distinctly different
from contemporary climates in the same areas and may even, in
certain cases, have been more arid than at the present day. It is
important to stress that the arid zone was not completely obliter-
ated during the pluvials. Reconstruction of the palaeogeography
of North Africa and the Middle East indicates that the Sahara
then existed as an arid zone, but in a restricted area with more

humid islands in the mountainous zones of Tibesti, Ahaggar and Ennedi. Much of the northern and southern Sahara had a hot steppe climate while the Sahel, the Sudan and the coastlands of Barbary and Cyrenaica had a humid climate. The 'Empty Quarter' of the Arabian peninsula was distinctly arid as it is today but steppe conditions embraced most of the land extending south-west from the valley of the Tigris and Euphrates. Much of Israel, Jordan and Syria and the high mountains of the Yemen then had a humid climate as did the contemporary rainshadow area of the Deccan of India. Similar high moisture provinces were present north of the Sea of Aral in what is now semi-arid country.

The return to the norm of aridity did not coincide exactly with the final disappearance of ice at the end of the last glacial but began at least 8,000 years before the end of the Würm, at a time when man had reached the Upper Palaeolithic stage of development in the Old World and had deserted the Sahara apart from the oases and river valleys so that the pattern of population distribution was somewhat similar to that of the present day. Since the Upper Palaeolithic there have been no climatic fluctuations on a scale comparable with those which took place earlier in the Pleistocene. In consequence their geomorphological effect has been much less. Humid fluctuations have, however, occurred during the Holocene, i.e. post-glacial times, and have altered the ecological characteristics of the arid lands by providing opportunities for men and animals to penetrate into areas which had previously offered little opportunity for survival. One of the most important of these humid periods which followed the arid period of the late-Pleistocene embraced about 3,000 years of Neolithic times and began in approximately 5000 B.C.

To this sub-pluvial of so-called Atlantic climate belong the rock pictures of the Sahara which depict game of the savanna type and the animals herded by the Neolithic nomads. Gazelles, antelopes and ostrich, rhinoceros, elephant and giraffe, hippopotamus and buffalo require not only good grazing but also adequate surface-water supplies. Moister conditions are also indicated by fossil tropical red earths, and by the record of the layers of sediments or stratigraphy; artesian wells functioned in the oases of Kharga and Dakhla west of the Nile leaving clay hummocks on the oasis floors to represent the sites of fossil springs. In the calcareous tufas on the sites of former springs have been discovered Neolithic artefacts. There is even pictorial representation of a savanna-like vegetation on the walls of tombs dating from the period of

the Old Kingdom of Egypt. It is significant that the Neolithic rock
paintings come from the highlands and the other evidence from
tracts marginal to the Nile Valley. For the Great Sand Sea of
Libya there is nothing to suggest that the Neolithic sub-pluvial
changed the basic aridity any more than the longer pluvials of the
Pleistocene inter-glacial periods. Evidence from south-west Asia,
India, Nevada and Lower California also seems to suggest a sub-
pluvial following the intensely arid climates of the late Pleistocene-
early Holocene and may possibly have been contemporaneous
with the Neolithic sub-pluvial of the Sahara. The spread of the
new agriculture of Neolithic times from its original hearths in
south-west Asia was obviously made easier by the moister climate
of Atlantic times and it was unfortunate that the moisture characteristics of the sub-Boreal climatic period which followed the
Atlantic should have reverted to that of pre-Boreal and Boreal
times with desiccation and less favourable ecological circumstances for the maintenance of the domain of the Neolithic
agricultural spread.

From the beginning of the sub-Boreal period (about 2500 B.C.)
the desiccation of the Old World arid zone is attested by archaeological and other evidence. Representations of big game of the
savanna type disappear from the walls of temples and tombs and
rock drawings. Instead of the open chase, the Kings of Egypt
hunted in enclosures; dunes invaded the Nile Valley from the
Western Desert and the floods of the Nile decreased in height,
indicating lower precipitation in the areas to the south. The Dead
Sea appears to have been about 8 m (25 ft) lower at this time and,
farther east, this period of drought appears to be linked with
some of the nomadic migrations of Mesopotamia. All the archaeological and geomorphological evidence points to the aridity of
Bronze Age times and the withdrawal of animal and human
populations to the water points afforded by oases and river valleys
with severe restrictions on the nomadic way of life.

By about 850 B.C., however, there was a fluctuation to more
humid conditions comparable with those of the present day and
for the last two thousand years, i.e. in sub-Atlantic times, there
have been but short-term fluctuations within a generally arid
climatic framework. It is in relation to this latter period that the
greatest controversies have arisen partly because the evidence does
not correlate easily from continent to continent and partly
because the period lies within the realm of history and the rise
of the great civilisations and empires of Greece and Rome in the

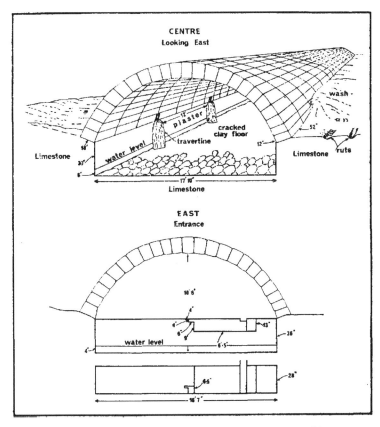

Fig. 2. Roman cistern at Saf-Saf, near Cyrene, North Africa

Old World. Was it increasing aridity or the breakdown of government and social institutions which led to the abandonment of their North African empire by the Romans and the decay of their great cities in the Levant? The writings of Ellsworth Huntington, which favoured desiccation, are not confirmed by the evidence of modern archaeology and stratigraphy. Everywhere within the bounds of the Roman Empire in North Africa there is ample evidence to indicate a preoccupation with water supplies. The great cistern at Saf-Saf near Cyrene in Cyrenaica (fig. 2) collected water from a small area of limestone pavement and was roofed with an arch of masonry to protect it against excessive evaporation. The underground aqueduct which leads from it to the city of Cyrene was constructed so as to reduce water loss to a minimum. In Syria the Roman 'limes' did not extend beyond the zone of permanent wells. Elsewhere in the Levant the remains of some Roman cities are situated in areas where the available moisture is sufficient to make them humid provinces at the present day, a fact which is concealed by the drastic deforestation and soil erosion which took place during and after the period of Roman occupation. In Middle Egypt later Roman pottery is found interbedded with aeolian sand and alluvium.

The evidence in North Africa, Egypt, the Levant and southwest Arabia points to the breakdown of civilisation with the collapse of the Roman Empire and its replacement by a culture which transformed the geomorphological and ecological balance. It was the replacement of the cultivator by the pastoralist, of the fields of wheat by the grazing sheep and goats, of the trees by grass or semi-arid scrub which gave but inefficient protection against the winter downpours that transformed the natural environment and the cultural landscape. The nomad penetrating deeper into the cultivated area had no use for the terraces and irrigation systems or the good soil of the wheat lands. Neither did the Arab of Cyrenaica who was content to let his flocks wander over the wheat fields of the Italian colonists of the fertile vale of Barce after the Second World War.

In North America there are bigger gaps in the story of climatic change, although the story is not complicated to the same extent as in the Old World by the actions of great civilisations until the arrival of the Spaniards. Using the evidence of the rise and fall of lake levels, the cutting and filling of arroyos, the formation of crusts and the microstratigraphic record, the record of the older dendochronological studies has been amplified and corrected. Tree-ring studies have been shown to represent mainly the length of the growing season rather than the amount of moisture received especially when based on the rings of the giant sequoia. The former lakes of the Carson Sink indicate a moisture maximum in the latter part of the last millennium B.C., but it was preceded and succeeded by drier periods when aggradation of the arroyos ceased and erosion was active.

The story of recent climatic fluctuations, say within the last hundred years, has been made easier for two reasons. The increase in interest in the earth sciences during the nineteenth century coincided with an extension of the scientific civilisations into the more arid areas of the world, notably in the USA, USSR, Africa, the Middle East and in Australia. Continuous climatic observations were made and supplemented by the gauging of rivers. The development of irrigation schemes as in India and Egypt and the south-west of the USA necessitated a much fuller knowledge not only of the contemporary water balance but also of fluctuations and trends. Accordingly there is an increasing mass of detailed knowledge available which demonstrates a lowering of the precipitation in lower latitudes while in higher latitudes there has been a warming up of the climate reflected in the retreat of glaciers, at least until the 1940s, and in the plant and animal ecology. The effects have been very striking in both the Old and the New Worlds. In North America it is linked with the Dust Bowl conditions of the 1930s; in Africa with the extension of the Sahara south into the Sudan and in the USSR with a lowering of the level of the Caspian Sea since the maximum recorded in 1882. The partial failure of the Russian virgin lands scheme may bear some relation to such climatic change or perhaps to short-term climatic variations. A fall in lake levels is reported also from East Africa and from Lake Titicaca in South America. From South Africa analysis of long-term climatic records indicates a downward trend in precipitation during the present century which is paralleled by the records for stations in the interior of Australia such as that at Alice Springs. In almost all cases the decrease in precipitation has taken place since the end of the nineteenth century but it is essential to note that the effects have not been uniform in every continent and that in some places there has been a positive tendency towards higher precipitation as on the tropical margins of the trade-wind deserts (the rise in the level of Lake Chad in the 1950s may be an indicator). Such regional variations are important in helping to decipher the anomalies of the climatic record inferred from the non-climatic data which is all that is available for the earlier periods of history and prehistory. In addition a knowledge of the fluctuations assists in the planning for better scientific utilisation of the arid zone. A knowledge of trends will assist in the avoidance of past costly errors of judgement in the economic utilisation of the semi-arid lands where schemes for reclamation and development are increasingly promulgated. Such projects are highly sensitive to climatic fluctuation so that an increase in the frequency of dry years would make them very vulnerable. It is interesting to speculate on the future of dryland utilisation if the climates of the Earth return to the norm of warmth and aridity which has prevailed for nine-tenths of the planet's existence of nearly six thousand million years.

3

THE DESERT LANDSCAPES

Aridity leaves an impressive mark on the earth's surface. Mountains and massifs tend to be discrete, slopes, where they exist, are steep, the relief tends to be angular although rounded forms also exist, there is much evidence of powerful sedimentation but a general absence of true soil, processes are intermittent and often directly related, as in sheet floods and mud flows, to heavy showers of rain at widely spaced and irregular intervals of time. These impressions of reality are, however, often clouded by romantic and fictional exaggerations as when Hollywood exploits in several epics a single dune near Stovepipe in Death Valley to give the impression of a universal sand spread in the desert lands when, in fact, less than 20 per cent of the arid zone is covered with sheets of sand or dune-fields. Even when the reality of the desert environment is perceived, its interpretation and a true understanding of its characteristics are made difficult by too rigid an application of the dictum of James Hutton that 'The present is the key to the past'. It is clear from the evidence for climatic change that the arid lands must be considered not only under the conditions of contemporary morphogenesis but in relation to those which have occurred earlier in geological time.

True aridity mummifies a landscape by slowing down the rates of change just as the dry sand and desiccating air preserved Egyptians and Peruvians for the museums of the twentieth century. As these mummies once had vitality, growth and change so had the arid zone landscapes in the past, with but few exceptions. Such landscapes betray the results of higher precipitation and reduced rates of evaporation during the 'pluvials' by the plethora of forms produced by the erosional and depositional processes of running water, chemical weathering and, in some cases, frost action. Modified but slightly by the heavy but sporadic showers of the present, the strange bedfellows of fluvial forms and the residue of savanna type weathering dominate a relief which was earlier believed to be the result of wind action. As in other morphogenetic regions there is a superimposition of land forms and soils derived from the amalgam of different bio-climatic systems of erosion and deposition. Such is the case when contemporary sand-dunes are seen to invade a Pleistocene river valley cut deeply into a plateau capped by a residual weathering crust. Here the weathering residues of a Pleistocene pluvial or Tertiary savanna climate have been preserved by Holocene aridity. It is unfortunate that the fossilising agents of the present have, until recently, been assumed to provide the active agent in landscape evolution in the past. Neither must aridity be assumed to be of equal intensity throughout the dry lands. In the extremely arid lands, using Meig's definition, relief and soils are preserved but on the semi-arid margins, where vegetation is less plentiful than in the better protected grassland steppe, morphogenesis is currently often powerful, even violent, especially where the spasmodic run-off is canalised in gullies or spreads out in sheet floods.

It is tempting in a geographer's assessment of the arid lands to select, for special attention, only those landforms which appear to have a positive potential and to neglect those which may be assumed to be of less significance or even negative potential. This would, however, give a false picture of the wide variety of habitats available within the arid zone especially if criteria more applicable to the humid zone were selected. Deposits of sand in northern Europe are generally neglected by the agriculturalist yet their equivalent, the 'erg' of the Sahara, may be of great biological significance. In areas of strong evaporation and low precipitation with high annual and seasonal variability, sandy surface deposits which allow easy and rapid infiltration of water can support persistent plant growth and provide grazing for the flocks and herds of the nomad. The rocky pavements, called 'hamadas' in the Sahara, may provide routeways for vehicles and animals or satellite launching bases, while the high bare massifs, such as the Ahaggar, now attract tourists and climbers in increasing numbers. Even the dry wadi assumes significance when its alluvial deposits are 'panned' for minerals or, perhaps more generally important, when it is realised that the stream floor sediments conserve humidity and ground-water flows. In fact, if one were to select environments of special significance within the arid zone, attention would be focussed on those areas which, by their evolution, have the capacity to conserve the most valuable natural resource of the desert—moisture. Where periodic or annual replenishment of moisture is possible as in the flood plains of allogenic streams or in the detritus-filled desert basins then conditions for land utilisation are at their optimum. Areas characterised by sedimentation must, therefore, receive special attention.

The processes of erosion and deposition in the arid lands conform to the bio-climatic pattern determined by the total environment—cultural as well as natural—and will vary as this environment changes in time and in space. This does not deny, however, the essential role of tectonics and lithology. The major deserts, especially those in lower latitudes, are often plains of vast extent corresponding to relatively stable areas of the Earth's crust subject principally to warping and epeirogenic (or vertical) movements. They are frequently margined by tracts of strong relief, by fault or fault-line escarpments or by mountains and dissected hill masses which result from differential tectonic movements or orogenesis. Within the desert plains there may be isolated hills or mountain ranges which, like the marginal highlands, provide the sources of rivers and make a substantial contribution to water supplies by acting as moisture islands of differing intensity with the changing climates of geological time. The volcanic hills of the desert lands perform a similar function to those uplifted in blocks between fault-lines or by elevation in the closing stages of an orogenic crisis. The local climate of a hill mass of whatever extent is different from the climate of the surrounding lowlands and, if of sufficient altitude and extent, may possess a regional humid climate as in the moist plateaux of Abyssinia which interrupt the arid tracts of adjacent Africa and Arabia. Conversely, the subsidence, down warping or rifting of a section of the crust tends to enhance the aridity of the depressed area as in the floors of the rift valleys of East Africa.

Tectonic movements may thus create anomalous moisture conditions within an arid zone and complicate the reconstruction of past climatic events. This is especially true of the Pleistocene period when many areas were subjected to regional uplift with local doming which rejuvenated the gentle relief of the planation surfaces created under different temperature and rainfall regimes during the Tertiary and early Quaternary eras. The increase in altitude following uplift must surely have caused some increase in rainfall which would have had important geomorphological and ecological consequences. The arid regions of North America, Asia and Africa are especially likely to have experienced these tectonic complications contemporaneously with changes in atmospheric circulation during the Quaternary.

Lithology is important in the arid, as in the humid zone, in determining the character of the landforms. According to the nature of the rock and its susceptibility to the special conditions of desert weathering, there will be variations in slope and in the type of surface deposits. These in turn will affect the local and microclimates and thus the ecological characteristics of the area and their utilisation by the sedentary cultivator or the nomad. Moreover the alteration mantle in an arid area will vary in its physical and chemical characteristics, assuming the same rock type, according to the climate which obtained during the critical period of its formation. It may possibly have been formed under the seasonally humid conditions of a savanna climate when the combination of rain at the season of high temperature proves a most effective setting for chemical weathering.

To the north of the low-latitude zonal deserts in North America and Eurasia the same rock will react differently from that of the savanna lands. In the temperate deserts and steppe lands, that winter cold and frost, which prevent the growth of the date palm, make a significant contribution to morphogenesis in the rate of rock disintegration. Compared to the hot deserts the modification of the rock proceeds much more rapidly under the effect of winter frosts since not only is there great humidity at times of thaw and snow melt which facilitates weathering processes but the formation of ice in crevices exerts its well-known shattering effect on the rock. In the most extreme examples, as in Mongolia, the debris and the forms it assumes are strongly reminiscent of sub-Arctic landforms developed under the periglacial system of erosion. Through frost action the rate of weathering of the higher zones within the hot deserts is increased compared with the lowland at the base although the effect is neither as great nor as prolonged as in the temperate deserts and steppes or higher latitudes.

Landform development within the arid zone at the present day thus varies with latitude and altitude. It also varies in a less easily defined way according to the intensity and frequency of the rain showers. The important showers, from both geomorphological

and ecological aspects, are those which exceed the infiltration
capacity of the ground and thus provoke surface run-off which
rarely has time to become concentrated and must perform its
function of erosion and transport in diffuse form. The heavier
showers are the more efficient erosive and transporting agents.
The greater their frequency the more they promote weathering
under conditions of alternate drying and wetting, while the
weathered product is quickly removed to expose a fresh unweathered surface. Unfortunately the occurrence of these extremely important showers has scarcely been systematically
observed and mapped in a way which would indicate regional
and local variations to the geomorphologist. It is known, of
course, that they are more common on the arid margins than in
the hearts of the desert zone and that they occurred more
frequently in the past than at the present day.

A few decades ago it would have been usual to include wind
action in this part of an introduction. Modern research has,
however, clearly demonstrated that, as in the more humid periods
of the past, the contemporary desert landscape is the work of
running water and that the wind plays only a relatively less important part. It is most effective when working on the old lacustrine deposits which it may erode, deflate or redistribute as long
as the grains are fine and incoherent. In fine debris the water-
table forms the effective limit to deflation. The fine dust is most
at the mercy of the wind; it envelops the coastal towns of
Cyrenaica when the winds blow strongly from the south and
has been collected far away from North Africa on the wings of
aircraft landing at Oslo airport; it is lifted by the turbulence of
miniature tornadoes, 'the dust devils', of desert and steppe the
world over. Yet areas of sand accumulation are only a small
proportion of the total area of the arid lands. The French Foreign
Legionary marched more over rock pavements and dry river beds
than over fields of sand-dunes which constitute less than 20 per
cent of the arid lands but gained prominence from their proximity
to the open corridors by which the desert caravans crossed the
Sahara.

The characteristics and evolution of desert landscapes

The traveller in the deserts of the Sahara, Arabia and Australia
accustomed to the relief of the world's humid lands is stunned by
the immensity of the vast plains which often extend unbroken to
the horizon or are broken forcefully by steep slopes. Even the

deserts which have experienced the most intense tectonic disloca-
tion such as those of the south-west USA or Central Asia are
dominated in area by lowlands with the mountains providing a
magnificent backdrop curtain. In the dry lands the structure is
clearly visible through the nakedness of rocks, while the marked
angularity so characteristic of the junction of slope elements
demonstrates the play of differential erosion and emphasises the
lithological contrasts. The hydrographic network is endoreic (see
Chapter 1, p. 14), terminating in closed depressions or basins of
interior drainage. The drainage is often disorganised and varies
in its degree of coordination according to the relief and amount
of precipitation now and in the past. Drainage basins in the arid
lands are usually smaller than in the semi-arid zone unless high
altitude intervenes, where a better organised drainage system, as
in the Ahaggar of the Sahara, is frequently found. Rivers are
normally ephemeral, like the showers which bring them into being,
so that the surface is crossed by dry river beds, the *wadis* of North
Africa and the Middle East, the *arroyos* of the American deserts
or the *nalas* of West Pakistan. In great contrast stand out the
allogenic streams which have their sources in the better watered
highlands of the arid zone or outside it altogether. It is the flood
plain of these perennial rivers, such as the Nile or the Colorado,
which offer the best habitat in the arid zone as density distribution
maps of population so clearly indicate. Apart from these linear
depressions of the river valleys, there are also the depressions of
the plains. Many are filled with alluvium or former lake sediments
which the wind picks up and re-deposits as a garland of dunes on
the leeward margin. Some are covered with glistening salt crusts
against which the green of the occasional salt-tolerant tamarisk
stands out amidst the bright reflected glare from bare rock, sand
and salt crystals.

The apparently flat plains of the great deserts turn out on
investigation to be made up of gently sloping surfaces of erosion
or deposition. Two main types of surfaces can be recognised, the
erosional surfaces of the *pediment* and the *pediplain*, both of which
may grade into surfaces of deposition. Genetically related to these
plains are the hills which rise abruptly above them, the *inselbergs*,
since the lower surfaces develop at the expense of the higher
ground as a wave-cut bench on the coastline extends, with cliff
retreat, at the expense of the hinterland (fig. 3).

The *pediment* is the hill-foot slope, a long, smooth erosion
surface leading away from the higher ground behind with a longi-

tudinal inclination which may vary from about $\frac{1}{2}°$ at the outer to as much as 7° at the inner margin. There is no lateral slope parallel to the hill front unless the pediment has been warped, so that the surface resembles that of a sloping lectern. Across this rock plane flow sub-parallel rills of water after the ephemeral showers. They do not incise their beds appreciably so there is no distinct valley and interfluve system. If a dissected pediment is encountered it denotes a climatic change and the pediment is thus a fossil form. Sometimes a veneer of alluvium is found on the rock surface, either continuous or in patches, which can support a sparse vegetation important for pastoralism but the pediment is essentially an erosional feature cut in solid rock. The junction with the higher ground may be in the form of a dihedral angle, a *Knick*, or by means of a short concave section of weathered debris which has fallen down from the steeper slopes above (30° in many cases). The junction may, however, be masked by alluvial fans at the mouths of gullies and larger valleys. These can coalesce to provide an almost continuous cover of gravels called *bajadas*. At the lower end, the pediment may grade down into a wadi or a surface of deposition where the rock plane becomes submerged beneath alluvial or lacustrine deposits in a depression.

Extension of the pediment takes place by the reduction of the higher ground through slope retreat. For structural and lithological reasons slope retreat may not be uniform and a series of embayments will be produced in the escarpment. If embayments from the opposing sides of the hill mass meet, then *pediment cols* or *passes* are produced. Further reduction of the hill mass can result in the almost complete coalescence of the pediments so that the landscape is made up of a series of low-angle surfaces dominated by residuals perched at their intersection. At this advanced stage of development a pediplain is produced which has the characteristics of combined pediments with *residual inselbergs*.

The problems of plains and inselbergs in the arid lands is, however, made more confusing by the intervention of climatic change. On the equatorial side of the hot deserts in the savanna lands there occur vast plains which have no appreciable slope. These smooth *savanna plains* are seasonally flooded during the period of summer rainfall. Watersheds are poorly defined and the direction of water flow may vary from year to year in braided stream courses which are scarcely incised below the surface level. Frequently they bear a deep residual weathered layer of decomposed rock since chemical weathering is very active during the hot

wet season. Like the pediplains they too are dominated by inselbergs which can be fringed by narrow pediments or rise directly
above the savanna plains.

The *inselberg* (*bornhardt* is synonymous) is a landform found
in a wide range of bio-climatic environments of the tropical zone
from the arid lands to the equatorial rain forests. It consists of
isolated strong relief with steep slopes and may vary in its dimen-
sions from a small isolated hill to a larger massif. German
geographers in fact differentiate between *Inselberg* and *Inselge-
birge*. The relative relief above the plain can range from a few
feet to several hundred feet. The shape varies widely; some are
domed, some have overhangs, others have a flat or undulating
top while others are asymmetrical. Lithology plays an important
part in determining their shape or even their very existence. Many
are formed of the same rock as the ground at their base, but others
are composed of more resistant rocks where the extension of
pediment or savanna plain was halted. It is not always easy to
distinguish between the two types since variations in rock resistance in apparently homogeneous material may escape even
detailed petrological analysis.

Desert weathering and erosion

Many theories have been put forward to explain these genetically
related landforms. They involve some knowledge of the weathering and erosion processes of the arid and semi-arid zones and
those of the marginal climatic regions, such as the Mediterranean
and savanna morphogenetic systems which are arid in the hot
and cold seasons respectively. As in the humid zones the role of
weathering is twofold—it softens up the rock to facilitate the
work of erosion and transport and works on the disintegration
fragments to reduce them in size and to provide the raw material
from which, with complex chemical and biological changes, the
soil is derived. Many of the processes of weathering, erosion and
transport under these conditions are not yet fully understood
although serious doubt has already been cast on some of the older
explanations.

It was understandable in view of the large diurnal temperature
range, and, in the temperate deserts, the marked thermal contrasts
between summer and winter, that mechanical disintegration of
the rocks should have received such prominence in the earlier
literature. Within the zone of penetration of the sun's heat and
effective annual and diurnal temperature variations (say to about

three feet below the surface) it was the expansion and contraction of the bare rock, poorly protected by an insulating layer of soil and vegetation, that was held to be responsible for the detachment of surface layers, for the cracks perpendicular to the rock surfaces and for the granular disintegration of the constituent minerals. Desert travellers reported, in fact, the 'pistol-like' cracks of exfoliating rock surfaces in the Sahara and Arabia. Rigorous laboratory experiments have shown, however, that thermal variations alone are ineffective and that moisture is required if the rock is to split instead of adjusting itself to the stresses and tensions which are produced. The presence of water or salt, both more commonly available in the deserts than formerly recognised, not only reinforces the thermal variations but provides the essential ingredients for chemical weathering which is thus far more active in the most extreme deserts than was previously supposed.

. Apart from the semi-arid margins with more frequent and effective precipitation, there are numerous sources of water in the desert zone. The occasional shower accompanied by the great heat affords an albeit short but effective chemical weathering environment which persists longer than the duration of the shower itself. Night dews, to which are attributed the vermiform markings on limestone fragments, are a substantial source of water for weathering purposes and are augmented by frost at higher altitudes and latitudes. If ground water is near the surface it may be drawn up by capillarity or water may be emitted by the rocks themselves. Ferro-silicic solutions exuded from the rock evaporate at the surface or in cracks where chemical weathering is more active. Clay minerals such as montmorillonite have been found in cracks (in, for instance, the Ahaggar of the Sahara or near Jeddah in Arabia). They take up more water during showers or dew and continue to provide moisture for chemical weathering. Shady crevices retain the moisture longest which probably accounts for the greater weathering on the shadow side of Egyptian monuments. A secondary effect of the exudation of mineral solutions is the formation of desert varnish which seems to act as a barrier to the weathering action of salt.

The action of salt in the arid zone is usually very effective. Wind-blown powder of sulphate and carbonate of soda deposited on crystalline rocks accumulate in the crevices and exerts tremendous pressure as the salt crystals increase in size. The power developed is sufficient to reduce granite to its component minerals which are removed by the impact of raindrops. Feldspars may undergo further chemical change by salt action. Sandstones are not immune to this form of attack although the rate of disintegration is less than that of crystalline rocks such as granite. The most resistant rocks are limestone and quartzite, which frequently form the high ground in the arid lands, while quartz veins stand out clearly in contrast to the rocks which they traverse. Granular disintegration through whatever cause is thus a prominent form of weathering and is responsible for the *taffoni* or honeycomb weathering which is prominent in many rock faces. Once started, the shade provided by the overhang accelerates the retention of moisture and the rate of weathering.

As the rainfall becomes more reliable towards the savanna and within the savanna itself, chemical weathering becomes much more effective. Not only is there greater heat and moisture combined but the vegetation becomes more dense and provides a greater quantity of organic acids to reinforce the mineral solutions of the soil water. Soil organic matter has also a greater power to retain water than the mineral fragments and thus plays a vital part in both physical and chemical weathering processes. The weathered layer of the humid tropics can reach a very great depth which, with the continuation of chemical and physical changes, produces the tropical red earths. In the savanna lands the breakdown proceeds more slowly and decomposition mantles are normally less thick than in the tropical rain-forest zone, but may still be of the order of tens of feet. Such soils normally show strong concentrations of oxides of iron and alumina and almost complete leaching of the silica. If the iron oxide has also been removed then *bauxite* is left but it is more common for the weathered material to be ferruginous. Ferruginous crusts, *ferricretes*, are found in the southern Sahara and Libyan desert and in Australia where they may be as much as 30 ft thick. They are hard and durable and form one of the duricrusts which may cap the landforms of the arid zone.

Other crusts are bonded by silica, e.g. the *silcretes* of South Africa, or by lime and gypsum, of which the *caliche* of Mexico and the south-west USA is a good example, although it is also found in Tunisia and other parts of North Africa. Calcareous crusts are normally quite shallow, extending not much over 3 ft in depth. Most frequently they develop on porous and permeable formations such as gravels and sands and can produce, as at Derna in Cyrenaica, calcareous sandstones from coastal dunes. In the lower section the lime forms a white powder either mixed with the formation or concentrated into rudimentary layers, but, towards the surface, it forms more compact and coherent crusts. Gypsum crusts are very similar in structure and both types may occur buried beneath mobile material. They occupy similar landforms, usually gently sloping alluvial surfaces such as terraces or plains. They appear to form as water enters the soil during infrequent rains and reacts with the soil particles. As the solution infiltrates it is subject to evaporation within the soil and precipitates the dissolved calcium carbonate or calcium sulphate. Under alternate conditions of solution and precipitation a thick *caliche* crust can be built up, the so-called 'cap rock' of the High Plains in Texas. The formation of crusts thus extends on the poleward side of the tropical deserts to form an important element in the environment of the arid lands. Saline crusts also cover large areas either in enclosed basins, such as that of the Quattara depression of the western desert of Egypt, the *schotts* of Algeria and Tunisia, the *sebkhas* of Libya, the salt pans of the Kalahari or the *playas* of the North American deserts or in many basins in Central Asia. They can also be formed through the misapplication of irrigation water and may be more effectively considered in that context.

Weathering, in association with the action of surface water, enters largely into an acceptable explanation of the formation of savanna plains and inselbergs which, as paleoforms, may now be found in arid and semi-arid lands. The thick zone of weathered material on the savanna plains varies in depth according to the lithology so that, below the regolith, the surface of the unweathered rock is uneven. During the wet season the rivers flow in braided courses and lack the ability to incise themselves deeply because of the gentle slopes and the fine-grained nature of the load derived by wash from the weathering residue. The rivers are able to cut laterally into the red soil surface, which offers little resistance to erosion, and very seldom encounter solid rock. The inselbergs within the savanna plains would therefore represent the uneven basement of the regolith which was exposed by the removal of the weathered layer under conditions of sheet wash and lateral corrasion. Once exposed the irregularities would be preserved and accentuated since the steep slopes could not retain moisture to the same extent as the ground at their base, and in consequence would be less subject to chemical weathering. At the same time the area of the plain is enlarged by the extension of pediments in the marginal higher ground as it retreats parallel to itself while the inselbergs on the plains may suffer a similar fate. The formation of pediments, however, appears to be more of a mechanical process in which much less chemical weathering is involved. It is thus more representative of semi-arid conditions. In view of the climatic changes which the arid lands have experienced it is not unreasonable to assume, however, that planation under savanna climate conditions is responsible for some of the major plains to be found in, for instance, Australia and North Africa since silicified wood of savanna vegetation has been found in the vast plateaux of the Sahara.

The vast *hamada* plateaux are bounded by wadis above which they hang with a marked cornice or angular junction. The plateau edge may be irregularly dissected by ephemeral streams whose valleys extend only a short distance into the plateaux where there is little evidence of stream action. On the plateau surfaces rain flows away without concentration into defined channels. The slope is so gentle, only 1:500 in the hamada of Guir in the Sahara, that the heaviest shower has little chance of forming concentrated run-off and evaporation and infiltration take a heavy toll of the precipitation. If, however, the surface is more steeply inclined then run-off becomes concentrated, incision takes place and the density of the wadi network increases. On the semi-arid margins with greater rainfall an even more perfect horizontal plane is required if concentration of surface run-off and consequent dissection is not to take place.

The formation of pediments has aroused some of the liveliest controversy concerning arid zone landforms. Few people would now claim that wind action was responsible but the protagonists of surface run-off and weathering are by no means unanimous as to the exact nature of the processes at work. Two major conditions have to be explained, the retreat of the back wall of the pediment and the plane surface of the pediment itself. The problem of scarp or inselberg retreat involves an appreciation of the origin of the steep slope in the first place. In the arid southwest of the USA strong tectonic relief provides almost ideal conditions as in the fault escarpments and monoclines of the Colorado Plateaux. The fault escarpments can reach tremendous proportions such as along the Grand Wash fault which forms the boundary between the Colorado Plateau and the Basin and Range province to the south-west. At its maximum the escarpment has a relative relief of 1,219 m (4,000 ft). The Hurricane fault, which extends from near Cedar City, Utah, for a distance of about 274 km (170 miles) to the Colorado River in the south, gives rise to the prominent Hurricane Ledge, an escarpment which varies in height from 30–427 m (100–1,400 ft) above the lower ground at its base. Monoclines are probably more representative structures than faults in this area and may create the steep slopes bordering plateaux and graben- and horst-like features. Other arid areas with retreating escarpments and pediments at their base lack such strong tectonic relief. For these King, for instance, postulates successive cycles of pediplanation developing from epeirogenic uplifts of continents and the incision of major trunk streams. Repeated rejuvenation with parallel retreat of valleyside slopes would produce extending pediments with broad plains at the base of the higher ground. It would appear doubtful, however, whether an area like the Sahara ever possessed major trunk streams so the explanation may not have world-wide validity.

The steep slopes of the regions with permanent or seasonal aridity contrast strongly with the gentler slopes of humid lands. They lack the mantle of weathered debris which masks the rock and by retaining moisture accelerates chemical weathering processes. In arid regions bare rock slopes are prominent and characterised by an upper section free from debris and vegetation —the 'free face', more or less vertical in section. From this face the weathering processes dislodge material which is disengaged in accordance with the varying rock types and picks out the microelements of the structure. Below the free face there is often an abrupt angular junction to the debris slope where material accumulates which has been dislodged from above. The debris zone terminates in another sharp angle above the rock plane at its base and is characterised by a greater capacity for the retention of moisture, sufficient to promote chemical weathering and sometimes enough for the growth of vegetation. Variations in climate and lithology can produce modifications of this basic layout. On the Mediterranean and savanna fringes the heavier seasonal rainfall induces greater chemical weathering which reduces the size and angularity of the fragments. These are washed down in the showers to form a short concave junction between the debris slope and the pediment. In temperate deserts with winter frost, shattering can have the same effect so that the North American or Central Asian deserts generally lack the abrupt Knick at the base of the slope. Lithology may also control the nature of the lower break of slope as when the disintegration of poorly cemented sandstones under arid weathering conditions provides a high proportion of fine material. Wind action and rain wash, of which the latter appears to be of much greater importance, remove the weathered debris from the slope and by a repetition of the processes the slope retreats maintaining its steepness. Some experts, however, stress that the slopes of the deserts of North America and the Sahara bear traces of much more active disintegration than is taking place at the present and, as usual, one is faced with the need to consider climatic change. This is especially important in the case of the crystalline domes which appear as inselbergs in the Sahara.

The action of water is now held to be responsible for the removal of the debris from the foot of the slope, which planes off the exposed rock surface left by the retreat of the slope and fashions the expanding pediment. Various theories have been put forward to explain the exact nature of this planation. McGee,[1] who was the first worker to recognise the pediment as a distinctive landform, favoured sheet floods as the main erosive tool but made little attempt to explain the nature of the backwearing process which produced the existence of a plane surface on which the sheet floods could operate. Others, such as Johnson, favoured lateral erosion by streams for both the formation of the rock plane and retreat of the back slope but his explanation is not convincing.

Forms of accumulation

According to de Martonne's calculation (see p. 14), nearly one-third of the land-surface is characterised by regions of interior drainage where the rivers terminate in enclosed basins isolated from the general base-level provided by the oceans. Many large watercourses flow into aggrading basins with a rising base level as long as climatic conditions remain unchanged. Since vast areas of the arid lands are composed of detritus-filled basins and valleys it is important to understand their characteristics. Not least of these is their property of water storage within the accumulations of alluvium protected from the strong evaporation which rapidly dissipates surface water. Such subterranean reservoirs have been used for millennia as a basis for pastoralism and cultivation in the arid zone and are, in consequence, areas of denser population. Such lacustrine or fluvial sediments may be redistributed by wind action and offer intriguing and distinctive forms of accumulation which, although occupying but a small proportion of the surface

[1] W. J. McGee, 'Sheetflood Erosion', *Bull. Geol. Soc. Am.* (1897), **8**, 87–112.

of the arid lands have an interest and importance which far outweighs their extent.

It is clear that organised drainage networks are generally relics of more humid past climates. Nevertheless, along these desert gutters flows the episodic run-off before it either infiltrates into the alluvium or is evaporated. In basins encircled by high land, the mountains which exclude the rain-bearing winds may trap sufficient moisture to sustain streams which penetrate for varying distances into the water-deprived basins. In both cases the character of the river channels and the regime depend on the pluvial characteristics of the source regions. Yet even the most dependable, such as those fed by snow melt or glacier meltwater, for example the Amu Darya and the Syr Darya of Central Asia, have a very variable discharge. Their transporting power is great during times of flood but for the remainder of the year they flow either in braided channels among their own flood debris or, more usually, do not flow at all.

At the change of slope between hill and lowland the transporting power of the streams is reduced with consequent deposition of the load. The coarser material is dropped first and the finer material is carried farther away from the hill-foot. The build up of arid and semi-arid alluvial fans is thus not materially different from those of more humid regions but the scale is usually far more impressive. In Central Asia spreads of fine-grained debris may extend for hundreds of miles from the hill-foot. Such far-travelled debris is probably the result of floods of some days' or even weeks' duration such as might be produced by snow and glacier melt in the spring and summer. Only streams fed by ground water can have a regular flow unless their sources lie outside the arid region. Allogenic streams such as the Nile or the Tigris–Euphrates or the Colorado which derive their discharge from beyond the arid margin are exceptional as are the more hospitable environments which they produce.

As a result of the intense evaporation and rapid infiltration only the finest debris can be carried for long distances together with the solution load; some of the fine debris is lost among the coarser elements of the river beds transforming the material into a loamy sand which may be suitable for seasonal cultivation—flood channel farming—or can support vegetation for long periods. Larger pebbles are abandoned as soon as they escape from the faster moving threads of water. The result is to give a heterogeneous accumulation of coarse sand and gravel which form shoals and islands with intervening bands of finer material. Vehicles have difficulty in crossing the finer-grained sections in such river valleys at times of low water.

In the lowest points of the basins may form lakes in which settle the finest sediments. In them also are precipitated the dissolved salts. Such *playas* or *sebkhas* are common enough in the arid zone and are genetically distinguished from the *schotts* in which water accumulates and evaporates from artesian sources. Vast salty lakes such as Lake Eyre in Australia or the Aral Sea in Central Asia have very variable depths and shorelines which recede under evaporation to disclose layers of saline sediments. These cracked clay plains, as in the eastern part of the Kara Kum, are important retainers of moisture compared with the very permeable sand spreads on their margins. In the Kara Kum artificial ditches have been constructed by caravans or nomadic pastoralists to lead the rainwater to pits. These wells offer a basis for livestock-rearing and, to allow the grazing animals to range in search of pasture, temporary reservoirs are constructed in the minor depressions of the main basin by Turkmenian herders.

When sodium chloride forms a large proportion of the precipitated salts, the cohesion of the plains is reduced through the flocculation of the clay particles. Wind-blown sand can then easily score furrows in the surface. Such *yardangs* may also be excavated by wind-borne grains in weak rocks other than clays. Sand, silt and clay particles deflated by the wind accumulate on the leeward margin of the depression to form a minor but readily demonstrable effect on wind action.

The work of the wind is not now regarded as a dominant factor in the development of the major erosional form of arid landscapes although it would be wrong to consider wind action as completely ineffective. On the margins of the arid lands it intensifies the results of the activities of man and his animals as factors in the formation of deserts. By deflation, by corrasion and by deposition the wind assumes an important role in morphogenesis.

Deflation works best with fine-grained debris which can be taken into suspension by the eddy turbulence of the airstreams. From the soils created in more humid periods of the past and from the fragments of contemporary rock decomposition, the finer material is removed leaving the coarser debris as the *reg*. In the northern Libyan desert the coarser particles, 2–3 mm in diameter, are found on the more exposed flanks of the shallow depressions in which the finer grained material temporarily

accumulates. The quartz grains are moved by saltation and surface creep, and at the most, are raised only a few feet above the
ground, depending on wind velocity and the roughness of the
ground surface. Saltation is encouraged on pebbly or rocky
surfaces and is restricted or inhibited by vegetation. Surface creep
seems to be most prevalent with higher wind speeds when a steady
forward progression of sand grains has been observed. Whether
abrasive wind action can be held responsible for the large hollows
such as occur in the Gobi desert in Asia is very doubtful. At most
it is now generally agreed that deflation can remove the alluvial
fill from tectonically created hollows.

Corrasion by wind-borne debris close to the ground can be
observed in weak materials and in wooden poles in desert lands.
It can also scour the paint from vehicles and houses but the effect
on resistant rocks may be to arrest or even prohibit weathering
and erosion. Sand blast has almost the same effect as the artificial
polishing of building stone, which by reducing the surface irregularities prevents the accumulation of moisture which form the
seat of weathering attacks. On the other hand, by removing
weathering fragments from *taffoni* of honeycomb weathering
hollows, it exposes fresh rock to the attack of the atmosphere. At
ground level, where wind-borne dust and sand attack are more
sustained, are found the well-known *Dreikanter* pebbles whose
facets seem to be indisputably shaped by wind corrasion.

Consideration of aeolian deposition leads us to safer but still
controversial ground. The great dune-fields, the *ergs* of the Sahara,
the Great Sand Sea of Salanscio in the Libyan desert, the *nefuds*
of the Arabian and the sands of the Great Australian deserts are
now believed, by many authorities, to be of quite local origin.
They would represent redistributed alluvial deposits which
accumulated in lake basins in the pluvial periods of the Quaternary. Undoubtedly some of the sand has been derived from the
breakdown of the sandstone rocks with which deserts such as the
Sahara are so richly endowed but much of this sand has probably
passed through the redistributed alluvial lake and stream stage.
For such vast dune-fields or *ergs* an enormous supply of sand is
necessary; when the sand supply is restricted, only thin sand
spreads or isolated dunes are found.

A typical thin sand spread, covering several thousand square
miles, is the *selima* of Libya where the bedrock lies just below the
sand surface. Under these circumstances it is possible for the sand
to be swept by the prevailing winds into dunes, margining corri-

dors of bare rock, the *gassis* of the western Sahara. These sand-free corridors have been long used by camel caravans and were crossed by the slave trains moving from the savanna lands and the tropical rain forest in the south to the desert oases and the Mediterranean coastlands. Modern motor transport sometimes comes to grief on these outcrops of bare rock and prefers the consolidated sand spreads where high speeds may be attained.

The complexity of dune forms within the great *ergs* of the Sahara and their equivalent in the other Old World deserts defy description in the space available here and it is only possible to indicate the major types. Important, however, is the fact that the Grand Erg Oriental and the Grand Erg Occidental of the Sahara, the *nefuds* of Arabia, and the *davans* of Afghanistan are the least mobile of all the sand spreads. Locally the sand may be endlessly redistributed but the basic forms remain as transverse or longitudinal dunes. The transverse dunes may be likened to huge ripples of sand with varying amplitude and wavelength. Whether the ripples are pronounced or scarcely defined, whether the distance between the crests is long or short, seems to be a function of wind velocity although, as in waves at sea, there is a limiting height to which the ripple crests can grow. Winds of great strength blow the grains of sand from the crest into the leeward hollow so that the amplitude of the relief is reduced. Elsewhere there are great longitudinal dunes which appear to be arranged in sympathy with the prevailing wind direction while the transverse dunes lie athwart the wind path. These longitudinal dunes may have very steep slopes with a knife-edge ridge from whose sabre-like appearance the name *sif* is often used in the Algerian Sahara. Under the influence of changes in wind direction they may locally be shaped into higher pyramidal or stellar dunes. These are frequently encountered on the margins of the *gassis* where they have the appearance of a cock's comb. Their steep slopes make movement from the sand-free corridors to the dune-fields difficult for animals and especially for vehicles, although heavy lorries have proved capable of ascending slopes 9 m (30 ft) in height at angles approaching 40°.

Apart from the *nefuds* of the Arabian desert, the finest examples of the major dune-fields undoubtedly occur in the Sahara (fig. 4) where they are separated by the outcrops of the bare rock platforms and the massifs such as Tibesti and the Ahaggar—in both of these massifs the surrounding sedimentary rocks arranged in the form of cuestas, and the crystalline basement, are dominated by the high volcanic peaks such as the Emi Koussi, 3,415 m (11,204 ft) of the Tibesti massif. One major dune-field extends from the coast of Senegal through Mauritania and is continued into Algeria as the Erg Chech which merges into the Grand Erg Occidental and the Grand Erg Oriental. Another is located on the margins of the Ahaggar massif and becomes the Sudanese *erg* in the south and the *erg* of Issaouane in the north, flanked by the *erg* of Murzuk in Libya. Farther to the east the Tibesti massif is flanked by the *ergs* of Ténéré and Bilma on the western side and on the east by the great expanse of the Libyan desert. All are affected and oriented in relation to the dominant north-east to south-west airstreams. Fossil dunes and sand-spreads extend far to the south to the northern limits of the Congo Basin where they are clothed with an open savanna woodland/tall grass vegetation. Farther to the west, in the middle Niger, many of the dunes are still mobile, but here it is believed that the biotic factor—the over-grazing by cattle brought to water in the Niger—may be the critical factor in the movement of the sand of some areas while other dunes remain fixed by vegetation.

In the popular mind there is little doubt that the crescentic dune or the *barkhan* is regarded as the principal dune form of the hot desert lands. Such *barkhans* are relatively rare in the Sahara but they are widely characteristic of the Arabian and Asian deserts and also occur in the deserts of North and South America, that in Death Valley being a particularly well-known example. These isolated dunes, in which the horns point downwind, seem to be related to areas where the supply of sand is intermittent and, in contrast to the great *ergs*, they are highly mobile discontinuous features, attaining great heights and towering above the flatness of the plains of sediments or rock on which they rest. Their gentle windwards slopes (often less than 4°) contrast strongly with the very steep slopes on the leeward side corresponding roughly to the angle of repose for dry sand (about 33°) while the horns and the abrupt break between the windward and leeward sides add to the distinctive appearance. Sometimes, as in the *nefud* of the central Nejd in Saudi Arabia, they may form on the surface of the larger dome-shaped dunes which themselves are margined by deflation hollows. On the other hand, in Mauritania and to the north of the bend of the Niger they are found in isolation. Isolated *barkhan* forms also occur to the south of the Tarim Basin in the Takla Makan where they dominated the desiccated clay soils of old lake basins. In Baluchistan there are well-known examples of

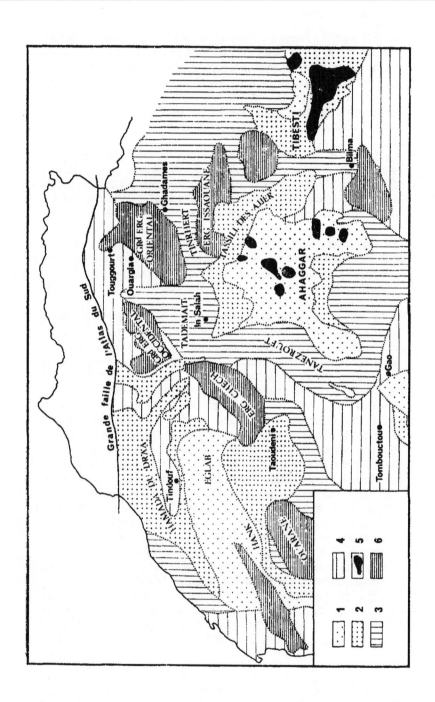

the *barkhans* 'marching' through passes in the hills where the winds are canalised but here, as in other areas where isolated dune forms are encountered, they seem to be related to the arid and semi-arid rather than to the extremely arid types of climates.

Dunes are not, of course, confined only to the inland deserts. They occur in the coastal deserts where the type appears to be mainly controlled by the character of the coastline and the wind régime. Low desert coasts, such as in parts of Peru and Bolivia, tend to be dominated by the *barkhan* form, but both elongated and transverse dunes are also present. All types may migrate inland for great distances until their further progress is restricted, as by the margins of the raised coastal plateaux in Peru, or by increasing humidity and more luxuriant vegetation.

Migrating isolated dunes impose severe problems on the communications of the arid zone, where the maintenance of economic activity is a function of the existence of routeways. Individual dunes can be crossed by tracked and wheeled vehicles fairly easily and prove less wearing on the suspension and the tyres than the rocky sections of the desert. On the other hand, there have been many accidents in the desert when inexperienced drivers have projected their vehicles over the break of slope between the windward and leeward sides of a *barkhan*. For railways the problem is comparable with that of snow clearance and protection in the sub-polar lands. Great difficulty was experienced in the construction and use of the Trans-Caspian railway from the Caspian Sea to Samarkand. In the first thirty miles from the Caspian, between the Merv (Mary) oasis and the Oxus (Amu Darya) and in the narrow belt between the Oxus and Bokhara, mobile dunes were countered by engineering devices which ranged from soaking the road bed with water from the Caspian to give it consistency, by covering sections with a layer of clay and by driving wooden stakes into the dunes on the windward side of the track against which the dunes could accumulate. Other attempts to stabilise

---

Fig. 4. Structural and surface features of the western section of the Sahara. (After Verlet, *Le Sahara,* 'Que Sais Je?', No. 766)

Key: 1 – crystalline basement
2 – sedimentary rocks of Primary age
3 – Secondary and Tertiary rocks of marine origin
4 – Tertiary and Quaternary rocks of continental origin
5 – volcanic massifs
6 – the great *Ergs*

the dunes necessitated the transplantation of tamarisk, from nurseries in the mountains of Persia to the sides of the road bed to stabilise the sand, but in spite of all these precautions it was still found necessary to have labour gangs available to keep the line open. In contrast, the construction of the railway between Biskra and Touggourt crossing the Grand Erg Oriental did not encounter insurmountable obstacles owing to the relative immobility of this huge dune-field, and there has been little difficulty in keeping the line open.

It is the *barkhan* rather than the *erg* which is the problem in the oases. At Jalo in Libya houses on the western margin have collapsed, as small mobile dunes within the oasis depression have migrated, but it is apparently unknown for the *ergs* to bury an oasis. This contrast between the major dune-fields and the individual migratory dune is thus important in many aspects of the geography of the deserts. The *erg* is a positive element of the natural environment through its water storage capacity, its ephemeral but relatively long-lasting pastures for the flocks of the nomad, its suitability for desert transportation by camel, road and rail and by the relative immunity from Pompeii-like interment of the oases which lie within its margins. Compared with the flood plains, the wadi channels and the marginal alluvial fans of desert basins, it offers a by no means insignificant habitat for the dweller in the arid zone.

Within the deserts of Inner Asia and especially on their eastern and south-eastern margins occur finer grained deposits than the sands of the Sahara and Arabia. Within China alone there are 259,307 sq km (119,000 sq miles) of *loess* although, of course, some of this occurs today within the humid rather than the arid and semi-arid zones. Loess is found in many topographic situations in Asia; sometimes it covers the mountain ridges and the intervening valleys, elsewhere it forms a thick sheet on fractured and dissected tablelands while on the lowland plains and piedmont basins it may be but feebly developed and contain large amounts of detritus and gravel. Although loess is generally described as a non-laminated, fine-grained deposit of Quaternary age, the deposits are by no means uniform. They vary in colour and in granulometry and may be stratified with intervening bands of pebbles. Their age has been variously interpreted as late-Pliocene or Quaternary and the diversity of views on the origin of the material derives, in part, from the variety of facies within the loess. At the base there is sometimes, but only in the valley bottoms, a torrential facies occurring as a basal conglomerate; in the central parts of depressions occur the fluvio-lacustrine facies (the stratified loess of Richthofen) which would represent the fine-grained material washed down from the slopes by rain into a playa depression from whence the stratification and the mixture of land and freshwater fauna. Bones, even complete skeletons, of rhinoceros, elephants, horses, oxen, deer, antelope, camel, etc., together with Palaeolithic quartzite implements (not of local origin), have been recovered from these rich fossiliferous horizons.

The typical yellow loess is regarded by many as representing an aeolian facies although its composition (both chemical and granulometric) is by no means uniform. It tends to have a very sandy component on the northern margins which reflects the proximity of the loose sand grains on the southern margins of the arid parts of Inner Asia. Farther to the south and east the silt or fine-grained fraction is higher except where fluvial action has led to a resorting and the introduction of gravel layers. In this yellow loess occur lime nodules either as separate concretions or arranged in bands. It would appear that earlier estimates of the thickness of the aeolian facies were excessive (Richthofen believed that in the steppe depressions it reaches a thickness of nearly 600 m (2,000 ft)), but modern estimates do not exceed about 400 m (1,300 ft). In many places, especially away from the Ordos Plateau, it is no thicker than 40–50 m (150 ft). There are good grounds for regarding the yellow loess as aeolian in origin but Russian and Chinese pedologists insist that true loess forms only through the action of edaphic processes and that the parent sediments may be of diverse origins. Loess has certainly accumulated and undergone the necessary pedogenesis within the prehistoric and historic periods. The dead in northern China are placed in thick wooden coffins left either on the ground surface or covered by a low tumulus. These have now been covered with loess deposits which are obviously of aeolian origin.

One is led to the conclusion that the evolution of the loess lands of Inner Asia and of the arid zone of the middle and upper Hwang-Ho mirrors the development of arid and semi-arid landscapes in the rest of the dry lands under the influence of fluctuating dry-wet climates in the late-Tertiary and Quaternary periods. In consequence all the morphogenetic processes, including the indirect effects of glaciation, have played a part with different intensities in time and space.

The soils of the arid lands

It is clear that for a great part of the arid zone, especially the extremely arid or true desert areas, the soils must be classed as azonal. The normal processes of pedogenesis and the development of horizons is constantly interrupted by the discontinuous nature of the processes of sedimentation and the winnowing effect of the wind while the lack of humidity inhibits the essential chemical processes from reaching their maximum development. There is also the absence, except in isolated instances, of a close vegetation cover and therefore poverty of organic matter. Under extremely arid conditions one can, therefore, expect to find true soils only in isolated places—on the larger scale within the oases (and not all of these) and, on the smaller scale, beneath isolated tufts of discontinuous scrub or bush vegetation. It is scarcely possible to refer, in the strict sense, to the 'soils' of the *hamadas*, the *reg* or the *erg* and some classifications of the 'soils' of the extremely arid regions are simply geomorphological divisions. Terms such as 'severely blown soils' refer to the *regs* or gravel deserts from which the fines have been removed by the wind or by slope wash. 'Deposited soil' covers the classifications of dune types on both micro- (*nebkhas*) and macro-scales (e.g. *barkhans* and *ergs*). In the 'deposited soil type' should also be included the 'medium' in which plants grow in many Saharan and Libyan oases.

From the pedological standpoint it is probably more correct to think in terms of 'skeletal soils' for the great part of the extremely arid lands. These are often divided into the 'lithosols', where the bedrock is resistant to weathering and carries little vegetation since the roots are unable to penetrate the solid rock unless it is well jointed, and the 'regosols'. The 'regosols' develop on more friable sedimentary rocks, on the deeply weathered igneous rocks (fossil survivals) and on the gravel elements in loess as in China and parts of Tunisia. These skeletal soils cannot be dismissed too summarily since the regosols may carry sufficient grass or shrub vegetation for ephemeral pasturage and have a high organic (rather than true humus) content. To these the term 'regosol ranker' has been applied.

In the less extreme arid and semi-arid lands but where, by definition, evaporation still exceeds precipitation, there is often a complete lack of leaching of the soluble mineral salts such as calcium carbonate and calcium sulphate. Indeed there is frequently an upward movement of soil moisture. Under such conditions the zonal soil is of the *pedocal* type (fig. 5) in which there are either nodules or layers of lime in the profile. In the most extreme types of *pedocal* the soil is composed almost entirely of mineral matter with less than 1 per cent of organic material. Here the calcium carbonate and sulphate may form a surface crust

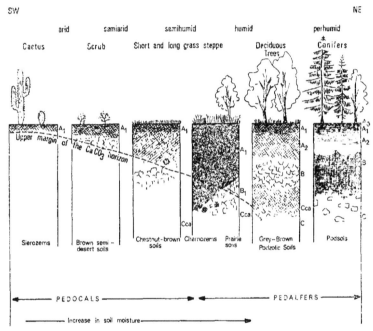

Fig. 5. Soil profiles representative of a cross-section of the USA from north-east to south-west. (After Thompson, *Soils and Fertility*, 1952, McGraw-Hill)

(e.g. the caliche) of lime and gypsum. The soil minerals are not normally available to plants when pedocalic conditions are excessive since the soil lacks the moisture which would make the nutrients available in a form which could be taken up by vegetation and cultivated crops. In fact, one of the functions of irrigation is to provide moisture for this purpose while dry-farming practices aim at conserving the water from showers of rain or at preventing the excessive evaporation of ground-water. Remarkable results are obtained in the cultivation of fruit trees, vegetables, fodder crops and even cotton in desert oases—of the deposited soil category—but in these the plant nutrients derive mainly from the minerals dissolved in the applied ground water. When the sand particles are too large, however, the full advantages of irrigation cannot be realised since, as occurs in the patches of more sophisticated cultivation in the extremely arid zone, the structure is too open for applied fertiliser to remain within reach of the plant roots.

On the margins of the very dry desert tracts, the rainfall is both greater in amount and is of seasonal rather than sporadic incidence. The superficial crusts of lime and gypsum may still occur but true soils lie beneath. These are the brown and grey desert and semi-desert soils called *sierozems*. Organic matter is still extremely scarce in these soils and is derived from shrubs such as *Artemisia*. As the rainfall increases with the increasing length of the rainy season, grassland begins to appear, interspersed at first with low shrubs and bushes. The higher humus content from the more luxuriant and more closely spaced vegetation gives a darker colour to the *chestnut-brown* soils of the steppe lands but the pedocalic characteristics remain, since the profile retains calcium carbonate. These soils, which occupy much of the drier part of the Great Plains of North America and the steppe lands of Asia provide, in comparison with the *sierozems*, a much more hazardous crop environment. Whereas the *sierozems*, and the climatic regimes in which they occur, are obviously unfavourable for cultivation, except under conditions where they are very fine-grained on the outer margins of moist flood-plains and alluvial fans, the *chestnut-brown* soils are situated in the transition zone where fluctuating periods of aridity and humidity tempt the farmer to extend the acreage under cereal cultivation in the wetter years only to suffer disastrous crop failures and 'dust bowl' erosion in the dry spells. Even in the cycles of 'wet' years, special cultivation techniques are required if high yields are to be obtained.

With greater humidity and more reliable seasonal incidence of rainfall, the grass cover becomes more continuous, more luxuriant and a better protection against shower erosion and the effects of sheet run-off. The organic matter in the soil is higher and under the influence of the greater humidity the processes of humification are better developed. Yet even in the *black earths* (*chernozems*) the humus content still amounts to only 10 per cent or less which is smaller than the dark colour of the 'A' horizon would suggest to the casual observer. Calcium carbonate, in nodules or layers, occurs in the *chernozems* profiles either at, or near, the base of the 'B' horizon. Many *chernozems* are derived from calcareous parent material but they are, of course, zonal soils, not intrazonal as this reference to specific parent material might suggest. With the *chernozems* one immediately associates the great cereal-growing areas of the world—the wheat of the United States and Canada, of the Ukraine and of Argentina. The greater rainfall reliability and amount define the *chernozem* zone as marginal to the dry lands of the world but since the *chernozems* are often developed from loess deposits in both the Old and New Worlds, aridity appears to be a major contributory factor in their formation.

Within the loess lands are also found soils which are similar to the *chernozems* but which lack the calcium carbonate concretions. These soils are obviously leached soils and the amount of leaching places this type—the *prairie soils*—into the humid rather than the arid zone. It is possible to underline the greater humidity of the *prairie soil* provinces by noting that these are the soils of the so-called Corn Belt of the USA. We have passed from the pedocals to the pedalfers although it must be recorded that the dry land margin as defined by the rainfall/evaporation ratio of unity does not coincide with the division between these two major soil groups. This accounts for the discrepancy mentioned in Chapter 2 where it was shown that the pedocals cover about 10 per cent more of the dry land surface than that defined in terms of the climate and vegetation of the present day.

With the exception of the true *chernozems* of Eurasia and the American continent, it is difficult to imagine more difficult zonal soils for colonisation by vegetation and for cultivation in the climatic zones where temperatures are high enough for the all-year-round growth of plant tissue. Over very large areas in North and South Africa, in the Americas, in Asia and in Australia cultivation is excluded from anything but the marginal steppe lands, the natural or man-made oases and the river basins. Yet just as it has been observed that there is no such thing as a desert mountain, because of the greater humidity induced by increasing altitude, so one must take into account the range of soil types found in the ascent from the interior of a desert basin to its mountainous rim or up the slopes of a desert massif rising above its surrounding plains. In such a transect might be encountered leached podzolic soils at altitudes which grade through prairie soils, *chernozems*, chestnut-brown soils and *sierozems* towards the floor of the desert basin under conditions where the true zonal

types can form (fig. 6). The marginal slopes of basins and the
flanks of desert massifs are thus a distinct possibility for grazing
and cultivation, especially if the natural drainage from the better
watered hill slopes can be diverted for irrigation purposes.

What then of the floors of the basins themselves? Here there
would appear to be distinct advantages for agriculture since there
is a combination of greater depth of sediments of fine-grained
composition at the termini of the drainage systems from surrounding uplands. Unfortunately, here also occur the main intrazonal soil types of the arid lands—the *halomorphic* soils—the
thick concentrations of soda, borax, lime and gypsum (and, of
course, the sodium chloride) of the *playas, sebkhas, schotts, vleis*
and the salt lakes of the western USA and Mexico, dry Africa,
Asia and Australia. Strictly speaking, deposits with such high salt
concentrations are not true soils and the term *halomorphic* is
normally reserved for saline sediments with a substantial silt and
clay fraction. Such saline soils occur wherever there is an excess
of the factors which produce the pedocal conditions; they are
poorly drained (hence the intrazonal classification) and are often
completely structureless in profile.

Within *playa* depressions the reason for the excessive accumulation of salts is obviously linked with the evaporation of surface
water but saline soils can also develop under other conditions and
in other sites. They may be induced or aggravated by the mismanagement of irrigation water and unscientific land-utilisation.
Saline ground water, found in many desert areas in the Old and
New Worlds and in Australia, may rise to the surface with an
increase in rainfall or through excessive application of irrigation
water (fig. 8). Lowering of the level of saline ground-water table
can also leave high levels of salt concentration in the soils within
the root zone of plants and cultivated crops. All saline soils offer
special problems for subsistence or commercial food production
which are not normally possible unless the salt can be flushed out
by carefully controlled application of irrigation water. Other
aspects of desert utilisation are also affected by high salt concentrations. They may either hinder or assist transportation in the
desert. With the great salt flat of the USA—the margins of the
Great Salt Lake in Utah—and more recently with the salt surface
of the fluctuating margins of Lake Eyre in Australia, are associated attempts on the world land speeds record. On the other hand,
the salt of the Quattara Depression, which extends below sea-
level in Egypt, concealed beneath an apparently firm covering of

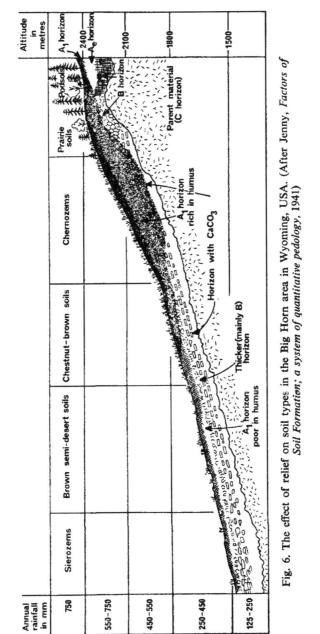

Fig. 6. The effect of relief on soil types in the Big Horn area in Wyoming, USA. (After Jenny, *Factors of Soil Formation; a system of quantitative pedology*, 1941)

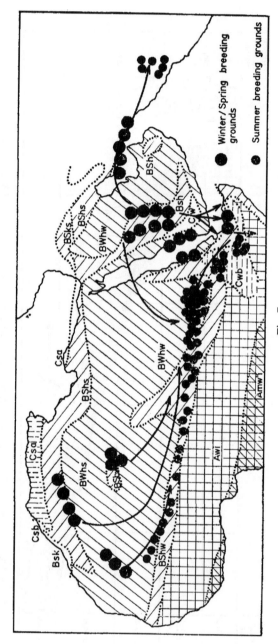

Fig. 7

clay, an obstacle to the mechanised movement of armies in the desert campaigns in North Africa during the Second World War. It was a critical environmental factor in the choice of the battlefield of El Alamein where the two flanks were protected—by the sea in the north and the impassable salt marsh depression in the south. No doubt the invention of ground skimming vehicles (hovercraft) will render strategic and tactical considerations of desert warfare rather more complicated.

Saline soils may be divided into two major categories, the white alkali (*solonchak*) and the black alkali (*solonetz*). These saline soils cover, in the USSR alone, more than 75 million hectares (30,000,000 acres) or 3·4 per cent of the land area. In the drier

Fig. 7. Winter and summer breeding grounds and migrations of the desert locust, 1968. (Based on Trewatha, *An Introduction to Climate*, McGraw-Hill, and information from Anti-Desert Locust Research Information Centre, London)

The remarkable coincidence of the 1968 distribution with the Köppen climatic zones is probably fortuitous but expresses the general picture

Key: Am – tropical rainforest climate
Aw – tropical savanna climate
BS – steppe climate
BW – desert climate
Cw – warm climate with dry winter (monsoon and upland savanna)
Cs – warm climate with dry summer (Mediterranean)
Additional abbreviations

A climates: w″ – two distinct rainfall maxima separated by two dry seasons.
i – range of temperature between warmest and coldest months less than 5°C (9°F).

B climates: h – average annual temperature over 18°C (64·4°F).
s – summer drought; at least three times as much rain in the wettest winter month as in the driest summer month.
w – winter drought; at least ten times as much rain in the wettest summer month as in the driest winter month.
k – average annual temperature under 18°C (64·4°F).

C climates: s – summer dry; at least three times as much rain in the wettest month of winter as in the driest month of summer, and the driest month of summer receives less than 30 mm (1·2 in).
a – hot summer; average temperature of the warmest month over 22°C (71·6F).
b – cool short summer; less than four months over 10°C (50°F).

continental climate of Central Europe there are over 500,000 salt-affected hectares (202,428 acres) in the Hungarian Plain alone; in China 20 million hectares (8,097,166 acres) have high degrees of soil salinity and there are many areas similarly affected in the Mediterranean lands of the Old World and the USA, in Australia and Africa and India and Pakistan. The *solonchak* of the arid Mediterranean lands or the southern steppes of Russia, which may be used as an illustration, is a white structureless soil with salty efflorescence but rich in carbonates which may be evidence that they are modified zonal soils of the *chernozem*, chestnut-brown and *sierozem* types. The pH values of the solonchaks vary from 7 to 8·5. Under conditions when excessive sodium chloride is removed and the sodium carbonates predominate, the black alkali (*solonetz*) soils occur; the dark colour derives from the solution of humus in the alkaline soil water. *Solonetz* soils also differ from the *solonchak* in that they have structured profiles, laminar in the upper layers and prismatic or columnar below. Leaching of the sodium carbonate, under conditions of higher rainfall or flushing by irrigation produces the *solod* (*soloth*) soils. They may be regarded as a naturally or artificially induced desalination.

Conditions of climate, terrain and soils are such that the dry lands appear to offer few opportunities for vegetation and cultivated crops. To the observer accustomed only to the humid lands, the prospects for the growth, reproduction and colonisation by vegetation under conditions of atmospheric and soil drought seem to be almost non-existent. Yet even in the extremely arid zones of skeletal soils there is often some vegetation which will vary in density and the area covered according to the variable rainfall amounts in time and space. As the humidity increases it has already been shown that the organic content of the soil increases, if not exactly in the same proportion, until finally true humus is encountered beneath the grass cover of the steppes. How does vegetation exist under these severe conditions of water scarcity and poor mineral or difficult saline soils? What are the special mechanisms of the vegetation to combat these problems imposed by the environment? In answering these questions we must consider the biogeography of the arid lands, firstly in relation to the vegetation and then in terms of the animal life which is a function of the vegetation. The vegetation, the herbivores and the carnivores reflect the total environments of the arid lands and will provide some clues to the utilisation of this dry third of the Earth's land surface.

4

THE BIOGEOGRAPHY OF THE ARID LANDS

There is little doubt that the total environment of the arid lands receives its clearest expression in the field of biogeography—in the way of life and reproduction, and number of species of vegetation and animals. This total environment includes the elements of climate, landforms and soils—each modifying the other and incapable of adequate interpretation unless considered as an indivisible whole. In this unity the role of vegetation is obviously of great significance. Firstly because it is part of the total environment and has such interdependent relationships with climate, geomorphology and pedology that to consider one without the other is meaningless; and, secondly, because all animal life and most human activity within the deserts is a function of the vegetation.

The ability of vegetation to modify the local climate has already been mentioned. It also has the role of protector of the ground surface from raindrop erosion; more actively, it binds soils and sediments together on the slopes to reduce the amount of mass movement. There is, however, probably no part of the total environment which can be so easily and quickly modified by short- and long-term changes in climate; there is no other element of the landscape which can be so easily changed by man and his animals; there is probably no other single component of environment which can cause such rapid changes in the appearance and potential of the landscape. Now because of the interdependence of the elements, a slight change in one component causes a chain reaction which, in the course of time, will lead to a reorganisation of the whole to produce temporary stability before a new series of changes is induced.

At the present day there is little that man cannot do in the way of large-scale 'geographical engineering'. One has only to think of the ambitious plans of Soviet engineers to maintain the level of the currently shrinking Caspian Sea; they aim to augment the contribution of the waters of the Volga by diverting the head-waters of the Vychegd and the Pechora, which flow to the Arctic Ocean, to the Kama, a tributary of the Volga. Attempts at rain making and urbanisation of arid lands fall into the same category. Yet for societies with a low level of technology, both now and in the past, it has been through the destruction and modification of the vegetation that men were able to induce, usually without conscious effort, morphogenetic crises and renewed erosion and deposition. This concept is applicable to all the bio-climatic zones but there are few areas of the Earth's surface where it is of such prime significance as in the dry lands.

From all these points of view the plant and animals of the desert merit detailed consideration. The more we know about the 'adaptations' of plants and animals to conditions of atmospheric and soil drought, the better we can produce strains of cultivated crops to suit conditions of aridity. The more detailed our knowledge of the physiology of the dry land animal population, the greater will be the possibilities for the success of large-scale stock-rearing in the drought lands or in areas of saline ground water. Perhaps there are now evolving even generations of mankind, types of men and women whose physiological functions differ from those of the humid zone. Is it true that 'the typical desert dweller has grey or light-blue eyes, and the muscles that control the pupils have been so developed that the pupils can remain as pinpoints without undue fatigue'[1] or would he prefer to wear Afrika Korps-type goggles if they were readily available? Not all the problems of life in the deserts and steppe lands have yet been satisfactorily answered but, with the increase in arid zone research stimulated and, in some cases, financed by the United Nations, the problems are at least more clearly formulated than they were before the beginning of the Second World War and already some applications of pure research are in hand.

There are numerous problems involved in the growth and reproduction of plants in the arid lands. They must survive under precarious conditions of moisture supply and have less oppor-

[1] Erle Stanley Gardner, *The Desert is Yours* (London, 1966), p. 37.

tunity to modify their local environment to suit their physiological requirements as the degree of aridity increases. Where the vegetation is sparse the plants lose the 'protection' of their neighbours; they become discrete individuals. The climatic component of the environment, in its most extreme form, is one of very high diurnal air and soil temperatures (within the root zone) and low nocturnal and cold-season minima (according to latitude and altitude) which impose severe strains on survival. Evapo-transpiration rates are high; the soil has usually a high mineral content and low organic content and lacks the moisture which would enable the plants to take up the available nutrients. The texture of the soil, to which the vegetation is very sensitive, is frequently open and may change rapidly over short distances as does the moisture content which is dependent on the pore space. In addition, some of the soils of the dry lands have a high salt content.

For the most part, then, the plants are *xerophytes*. They have the ability to arrange their life-cycles to suit drought conditions of varying length in areas of low relative humidity and high rates of evapo-transpiration in soils of the pedocal type with little moisture. The sub-soil within the range of the plant roots is almost permanently dry. To survive under these conditions of *xerophily*, plants, which are assumed to have had their origin in more humid areas, must make certain adjustments by developing mechanisms to resist heat and drought. Not all the plants of the arid lands are *xerophytes* however. Some survive because they are able to evade the drought conditions, others because they are able to evade the high salt concentrations of some dry land soils. *Xerophytes*, drought and salinity evaders still have one major enemy with which to contend—the animals which graze on them and the people who burn them for fuel and clear them for cultivation. To these hostile desert-dwellers the plants make a characteristic thorny or poisonous defence.

The drought-evading plants are not true xerophytes since they adapt their life-cycles to the periods when moisture is available. Some desert plants have exceedingly short life-cycles, e.g. the *Boerhavia repens* of the southern fringe of the Sahara which has been observed to flower, die and sow its seeds in the space of eight days when moisture from an ephemeral shower is available—i.e. under *mesophytic* rather than xerophytic conditions. Such ephemerals are annual plants and grasses found typically in the less extremely arid climates with winter or summer seasonal rainfall. Such winter or summer annuals are found in the Mojave and

Colorado deserts of the USA and on the northern and southern fringes of the Sahara. The seeds lie dormant after dispersal, sometimes through the next period of rains, which gives rise to the concept that the seeds have built-in inhibitors which prevent germination until the chances of achieving the full life-cycle are certain. Others such as the salt-bushes (genus *Atriplex*) disperse two kinds of seed which germinate after different intervals of time, moisture and temperature conditions. Drought-evaders also make economical use of the limited moisture supply from the soil. They have low water requirements (the most successful cereals of the dry lands combine low water consumption with a short life-cycle) and are widely separated to avoid competition for soil water. This wide separation, common in the drier areas and so characteristic of the deserts, may give way to dense patches with a greater number of species where a more plentiful supply of water is locally available as in the floors of wadis. Unfortunately these are also the breeding grounds of the desert locust. The drought-evaders also have root systems which are large in proportion to the stems and foliage while the individual leaves are very small— in fact the evaporating mechanism of the leaf may be replaced by a spine which also acts as a deterrent, but not a complete defence, against the browsing animals.

The grazing camels, antelopes, gazelles, goats or sheep are used by the vegetation to assist in the dispersal of seeds over the widest possible area to provide the best chance of finding good environmental conditions for germination. Barbs, burrs and bristles on the seeds and fruits are easily attached to the legs and coats of animals and it has even been suggested that some forms of vegetation (chenopodiacae of the genus *Kochie*) previously unknown, but now growing in profusion on the margins of the Libyan desert, were carried there on the boots of Australian soldiers in the Second World War. The tendency for desert plants to possess mechanisms for long-distance seed dispersal is exemplified by the rolling habit of some fruits and plants such as the 'Rose of Jericho' (*Anastatica hierochuntica* and *Astericus pygmaeus*). Other methods of seed dispersal, in which the wind plays an important contribution, offer similar interesting methods for survival and reproduction under the most favourable conditions. Think of the tumble-weed rolling through the streets of a ghost-town in the American south-west in a Hollywood film.

The *xerophytes* proper—trees, plants, perennial grasses— possess two main mechanisms to combat drought. Some are *drought-tolerant*; they can survive long periods of dehydration, like the creosote bush (*Larrea divaricata*) of the North American drought lands, by passing through low moisture periods in the vegetative stage. Growth continues when water is again available although, in the interim, the plants look dead and brown. Many fungi, lichens and mosses are drought-tolerant but comparatively few desert plants fall into this category. Other *xerophytes* have developed *drought-resistance mechanisms*. They reduce transpiration by dense, hairy coverings of the leaves, by closure of the stomata and by shedding leaves at the beginning of the dry season. Others possess the same structures for reducing water loss but may also impound water in the leaves, roots or stems. These are the desert *succulents*—the cactus and the prickly pear, for instance.

Avoidance, tolerance and resistance then are characteristic of *xerophily* but vegetation in the arid lands must also be able to cope with the high salt concentrations of the *solonchaks* or the *solonetz* intrazonal soils. Many *xerophytes* have the ability to resist the toxic effects of salinity in soils and ground water. They have developed tolerance mechanisms so that they may survive with high salt concentrations in the cell sap; some can exclude salt or actively excrete excessively high salt concentrations; others evade the salt damage by regulating the life-cycle to the period when there is a high moisture content in the soil and thus a low salt concentration. The response of such vegetation to conditions of high salinity is obviously of interest to the farmer in the arid lands.

This 'adaptation to drought' of the vegetation of the arid lands is thus an extremely complicated combination of physiology and anatomy which the simplified summary above does no more than suggest. A general description of the vegetation would state that the plant life is poor in species and consists of widely separated plants which grow closer together as moisture increases on a seasonal or annual basis. To counter soil and atmospheric drought the leaf surface is small in comparison with the root surface; the roots penetrate deeply and explore laterally to seek moisture. The stems are small and often protected by a thick cuticle or cork bark while the stomata are sunken or protected to reduce transpiration. In the most extreme deserts the vegetation consists of isolated tufts many feet apart. Towards the transitional margins of greater humidity the grass begins to appear regularly with the seasonal rains—one is moving out of the true desert into the steppe or pseudo-steppe, into that large part of the arid lands, probably four-fifths of the whole, which is loosely termed desert but which is really steppe (cf. the North American 'desert'). Obviously wide liberties are taken with these terms, and bio-geographers and plant ecologists are among those scientists for whom indices of aridity—or quantitative measurements of drought—are prerequisites for study of the arid zone.

Some arid lands have clearly defined types of steppe vegetation. In the Sahara, the Chilean desert and the Nefud of Arabia, there are zones of extreme aridity, the true desert, which at the present day form a clear if not continuous break between the vegetation of the summer and winter rainfall margins. This break is not so apparent in the dry lands of the American south-west. There the sage-brush steppe in the north mingles in the intermediate zone with the mimosa steppe in the south, although farther south there is greater dominance of the cactus and the agave which offer much less favourable forage for animals. The spineless cactus which has been developed through selective breeding is still more of a laboratory plant than vegetation in its natural habitat.

Animal life

The seed-dispersal mechanisms of some arid-land plants demon-strate the need for mobility to ensure survival of the species. For the animal kingdom, this mobility, by running or jumping, is expressed in its most extreme form by the gazelle (the pursuit of which was to promote the first real attempts to use motor-vehicles in the extreme and semi-deserts), by the antelope and by the kangaroo. These large mammals are, however, relatively scarce in the desert areas and are more commonly found in the semi-arid lands. They survive because they have the speed to reach the water-holes and to take advantage of the pasture which becomes available after the rains. With varying degrees of success these animals have been hunted by man since prehistoric times, as is indicated by the rock drawings of Tibesti, but hunting could not provide a livelihood for a large number of desert-dwellers even when man had learnt to arm himself with slings and arrows to compensate for his lack of speed and endurance over long dis-tances. Fortunately there were also slower moving mammals— the walkers—which he was able to control and domesticate. These animals, such as the camels, sheep, goats and cattle possess mechanisms for survival under drought conditions although they must take in moisture from springs or from vegetation at regular intervals. Without such herbivores dependent on the vegetation there could have been no development of the nomadic way of life, no symbiosis between the desert and the oasis, no struggle between the steppe and the sown. It is also significant that the desert and the desert-scrub lands lack major predators in quantity since there is insufficient food available for large carnivores. Nomadic stock raising would have been less possible if there had been many lions, leopards and pumas to prey on the flocks.

The runners and the jumpers of the desert have the ability to seek water over long distances; they are thus less representative of animal adaptation to drought than the slower-moving large and small vertebrates and invertebrates. Small mammals, birds, insects and reptiles depend on, and must adapt themselves to, their local environment and were it not for their involvement in the total ecosystem and the mechanisms of adaptation to drought which they demonstrate, there would be, apparently, no real justification for discussion at length in a geographical essay. Yet to this category also belong the ubiquitous desert flies, which are really denizens of the oasis rather than the desert in the strict sense. In the oases they rise in swarms from the dates drying on the flat-topped roofs of the houses and accompany the desert traveller on his journeys from oasis to oasis. The problems of health to which they give rise are probably equalled by the discomfort imposed on the European newly arrived in the desert, or on the people of Egypt—'and there came a grievous swarm of flies into the house of Pharaoh, and into his servants' houses, and into all the land of Egypt: the land was corrupted by reason of the swarm of flies' (Exodus viii, 24). Reptiles, found even in the most extreme deserts, may be poisonous to men and stock although perhaps less dangerous than is commonly supposed, while the diversions of the lizards' antics on the walls of houses offer a humorous relief to the stringencies of traditional desert life. Disney has shown that even the scorpion can be made into a figure of fun. The locust, in contrast, is without doubt the greatest menace since it is the consumer of green stuff and by its mobility carries destruction into the cultivated margins of the dry lands.

In crevices in the rocks, in caves and where animals can dig into regosols are encountered the first type of adjustment to the dry land environment—the principal of drought and heat evasion. Burrowing animals are widespread in the arid zone; they eat roots and insects and are represented by the jerboa (or desert rat), moles, rabbits and even birds. By burrowing they can create their own micro-climate away from the high diurnal air and upper 'soil' temperatures. Animals such as the kangaroo rats and jerboas are also nocturnal. The day is passed below ground in the micro-environment of their burrows where the humidity can be up to five times as great as the air outside. This diurnal/nocturnal rhythm of life is paralleled in some species by adaptation to seasonal conditions. Just as some animals of polar and severe continental climates hibernate, so some hot dry climate animals *aestivate* during the hottest season of the year to reduce body temperature, rate of respiration, transpiration and the need for food and water intake. They become dormant rather than quiescent and emerge only when temperature and moisture conditions are particularly suited to reproduction and completion of the life-cycle. The corollary of these adaptations is, of course, that the animal life visible in the dry lands will vary from day to night and from hot season to cool or cold season.

Earlier it was noted that some species of vegetation counter conditions of drought and heat by providing themselves with thick cuticles to reduce transpiration. Similarly there are many forms of animal life in the arid zone which have acquired comparable mechanisms—the development of a shell, as in the desert snail or the wool of the Merino sheep. Fur, wool or hair helps to insulate the animal from high air temperatures and reduces transpiration. Some plants were also shown to have short life-cycles to take advantage of temporarily favourable environmental conditions; in the animal world this characteristic is possessed by bees, wasps, hornets and spiders, and the locust also depends on conditions of temporary advantage. In the uncultivated areas locusts feed on the grasses which follow the rains and to a lesser extent on the scrub vegetation. Since a 'plague' of locusts has always had disastrous effects for the use of the dry lands of the Old World and Australia their habits must be studied in more detail.

The locust provides a good example of the religious sublimation of economic necessity. After the green crops had been devastated famine could be avoided only by eating the locusts themselves. The Old Testament, in distinguishing between clean and unclean meats, states: 'Even these of them ye may eat; the locust after his kind, and the bald locust after his kind, and the beetle after his kind, and the grasshopper after his kind' (Leviticus xi, 22). No doubt the addition of honey made them more palatable for John the Baptist (Matthew iii, 4).

The locust is physiologically not well adapted to drought conditions—it is a drought-evader which requires moisture for its short life-cycle and it survives by its mobility. It is, perhaps, paradoxical that attempts to provide soil moisture by irrigation have increased the potential breeding grounds of the locust swarms and reduction of crop yields which irrigation was designed to increase. Like the mosquito, which is an unwelcome cohabitant with man of the oases, the locust moves downwind and because of its poor physiological adaptation to drought is strictly under the control of the moisture regime. Locust swarms are very sensitive to rainfall variability, to wind direction and frequency and in the past the visitation of a swarm of locusts was entirely determined by meteorological conditions. Some areas were devastated, some escaped according to the distribution of the low-pressure systems which produced the rain and the grass, and towards which the in-blowing winds brought the swarms. In Africa, however, two of the most dangerous types, the Red Locust and the African Migratory Locust, have been controlled since their breeding grounds were located and measures taken to prevent the breakout of swarms.

In contrast, the desert locust (*Scistoceria gregaria*) has proved less easy to control. In the last twenty years there have been recurrent locust plagues across the Old World deserts from the Atlantic to the Indian Ocean. Information from the Anti-Locust Research Centre, published as a monthly summary, indicates that the Desert Locust breeds in the zone moistened by the late winter/early spring rains of the Mediterranean margins of the Old World deserts. At this time spring up the ephemeral grasses on which the locusts feed. The females then deposit their eggs, which have high humidity requirements, preferably in surface-dry sand over a moister layer beneath. This breeding zone thus coincides with the late winter/early spring shower belt of North Africa, the shores of the Red Sea and northern Arabia through Persia to Pakistan and north-west India. In the summer the swarms bred in these northern areas migrate south across the Sahara and move towards the east into Pakistan. On biogeographical grounds one would expect another feeding and breeding zone associated with the summer rainfall margins of the extremely arid regions and this indeed occurs—a belt across the Sahel and the Sudan extends into the southern Arabian peninsula and on to Pakistan. Between these main belts the locusts migrate, controlled by the seasonal occurrence of rainfall on the Mediterranean and savanna margins (fig. 7). Disruptions of this breeding and migration pattern may occur as the result of the high rainfall variability of these regions. This pattern is complicated today by insecticide control in the breeding grounds. Comparitive studies in geography, both in place and time, often indicate that a remedy to an environment/ land-use problem is often initiated only as a result of a major economic or natural catastrophe, or an interaction between them, in a country or area which can invoke cooperation and economic resources in proportion to the damage which has been done. In the present context, in the winter 1954–5, swarms of locusts destroyed £4½ million of citrus fruit trees in Morocco. By the late 'fifties and the early 'sixties, air-borne insecticides were sprayed on a carefully coordinated ground/air control programme; in one season, 1959–60, over 3,400 tons of insecticide were used against the winter breeding grounds in Morocco and extended, as could easily be arranged under the political framework of north and west Africa at that time, over the Sahara into the French controlled or economically associated territories in the south.

Towards the east, political fragmentation and more pronounced under-development hindered the introduction of similar measures until, in 1962, the Food and Agriculture Organisation of the United Nations sponsored a sustained and expensive attack (£3½ million to begin with) which included research into the ecology of the locust whose normal habitat, as mentioned earlier, coincides with areas of deposition where the moisture content is sufficient to support a scrub vegetation at least, i.e. wadi floors, inland drainage foci, alluvial fans or perennial and ephemeral streams and flood plains. These are the intrazonal components of the extremely arid lands which provide ecological islands and routeways and which also offer the best opportunities for pastoralism, sedentary subsistence or commercial crop farming. The mobility of the locust allows the swarm to move from one ecological island to another in a direction determined by wind and rainfall. It is this very mobility which demands international cooperation just as a dry land perennial river requires international cooperation if it is to be used successfully for purposes of irrigation.

Under the stress of the plague, this cooperation against the locust has to some extent been achieved. Yet by the mid-1960s the FAO were issuing new warnings of swarming locusts in Iran, Pakistan and the Atlas mountains. The subsistence cultivator or pastoralist or the small peasant farmer in Persia, West Pakistan

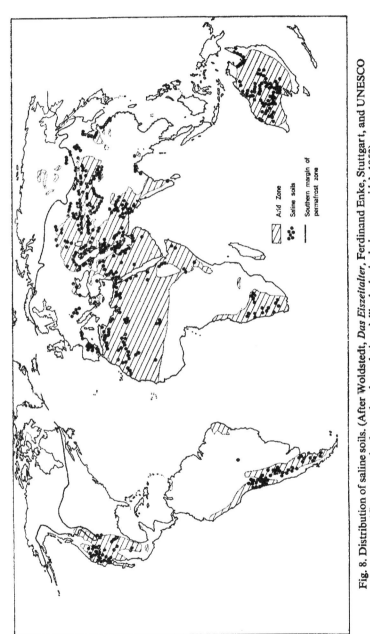

Fig. 8. Distribution of saline soils. (After Woldstedt, *Das Eiszeitalter*, Ferdinand Enke, Stuttgart, and UNESCO 'Compte rendu des recherches relatives à l'hydrologie de la zone aride', 1952)

Arid Zone

Saline soils

Southern margin of permafrost zone

or the dry zone of India may now receive a measure of protection from famine and death as a result of measures taken to ensure the continuity of commercial citrus fruit production thousands of miles away in Morocco. In the distribution and migration of the desert locust, insecticides, applied from the air or on the ground, thus appear to have reduced the environmental control of the east wind called for by Moses and the 'mighty strong west wind, which took away the locusts and cast them into the Red Sea' (Exodus x, 19).

If the locust represents a major hazard in the history of dry land utilisation in the Old World, yet is not of the desert, it is fortunate that the arid lands also contain a beneficent animal population whose physiological adaptations permitted utilisation in a way, and at a stage in technical progress, which would not otherwise have been possible. How could the arid lands of the Old World support a population, not primarily based on oasis cultivation, without the camel? Even the city-dweller was dependent on the camel caravan for the trade which supported the desert cities. What would have been the opportunity for subsistence or commercial pastoralism towards the more humid margins without the humped Zebu cattle and the fat-tailed sheep? This is not to underestimate the importance of the horse in the desert but it is essentially a grassland animal which must carry its own fodder into the extremely arid lands and be regularly watered. The ass and the mule also figure prominently as work animals in the desert and are better suited than the horse to these conditions; the ox is also a most valuable beast of burden and is able to carry heavy loads for three days without food or water.

One might ask why knowledge of the physiology of these animals was so long delayed once the technically advanced civilisation had begun to explore the desert. A better understanding of the 'ship of the desert' might have saved the lives of Burke and Wills in their exploration of Central Australia. In 1846, with camels as transport animals, they were returning to Cooper's Creek from the Gulf of Carpentaria when Burke and Wills succumbed to heat exhaustion promoted largely by mismanagement of the camels on which their safety depended. It was not then realised that imported camels from the Old World required two to three years' acclimatisation before they could be used in the arid lands of Australia; the journey by Burke and Wills was made during the early stages of acclimatisation.

Acclimatisation is obviously a critical factor in the successful introduction of grazing animals into arid lands but it is also important to have some previous knowledge of their resistance to drought. Apart from the jerboas and other small mammals, insects and reptiles, there is really only the camel to provide a norm against which the other herbivores may be assessed. For man the camel is the most vital animal in the arid lands of the Old World and it was the lack of a similar animal in the New World which delayed the exploration and exploitation of the arid south-west of the USA and the dry area of South America. This does not mean that the camel is ideal as a beast of burden or as a provider of milk, meat and leather. Of the two types of camel, the Bactrian with two humps derives from Central Asia and can withstand greater cold and damp; it can also go longer without water than the selectively bred one-humped dromedary of the Sahara and Arabia. Explorers' reports give widely varying figures on endurance, some of which will be resolved later, but many maintain that the maximum endurance of the dromedary without water is about ten days while the Bactrian can go for up to thirty. These figures must be treated with reserve. Numerous complications are concealed within these simple statements, including the season of the year, the kind of fodder available to the animals and the amount of time they were actually worked.

Neglecting its objectionable habits, one obvious characteristic of the camel, shared with other pack, draught and riding animals, is that it consumes large quantities of fodder—18–32 kg (40–70 lb) a day—and fodder is not readily available in the desert. This demand for fodder controls on the one hand the itinerary of the traveller since pastures must be sought, often away from the most direct route, and on the other, restricts the numbers of beasts which can be employed in a caravan or grazed by a nomad to the size of the pastures available. Feeding also reduces the effective working time of the camel. Added to these delays imposed by nutritional requirements is the fact that the camel is a slow mover; its daily journey under the best conditions is not more than about 32 km (twenty miles). Admittedly there are racing camels which are specially trained to cover up to 65 km (forty miles) a day, but they are too few and too valuable to be used as pack animals in caravans. The camel is thus a slow-moving animal which requires frequent rests, yet it can carry up to about 152 kg (3½ cwt) which is considerably more than the yak, llama or sheep used in other desert areas. This, however, does not make it into a bulk carrier of cheap or inessential articles; on economic grounds it should carry luxury articles of high value. Yet in Australia and in Pakistan it may be used to draw a cart, or, in some areas, the plough.

It is able to perform these duties by reason of its anatomy and physiology and herein lies its main advantage. It has large feet which spread the body weight (say half a ton) over a larger area of soft sand than the smaller hoof of the horse; its pads are thick to provide insulation against the extremely high rock and sand temperature and, of course, it has a 'hump'. Explanation of the camel's hump(s), or the hump of the Zebu cattle or the fat-tail of the desert sheep leads us from anatomy to physiology and to the only recently understood functions of these portions of the anatomy. With this knowledge it is now possible to answer the main question concerning the camel—for how long can it go without water and why is it able to endure much longer than man under the same environmental controls?

McGee has provided a detailed account of a human being in an advanced stage of water deprivation, that is without water for one or two days in the heat of the summer in the south-western 'deserts' of North America, or for up to one week in a cooler climate, assuming that there has been no intake of moisture from juicy fruits such as oranges, tomatoes or melons. 'He was stark naked; his formerly full-muscled legs and arms were shrunken and scrawny . . . his lips had disappeared as if amputated, leaving low edges of blackened tissue; his eyes were set in winkless stare, with surrounding skin so contracted as to expose the conjunctiva; . . . his face was dark as a negro and his skin generally turned a ghastly purplish yet ashen gray . . . his lower legs and feet, with forearms and hands, were torn and scratched by contact with thorns and sharp rocks . . . yet without trace of blood or serum.'[1] This man exhibited all the signs of an advanced stage in body moisture deficit and understanding of the behaviour of the human body under such conditions of water deprivation helps in the comprehension of the heat resistance and low water using mechanisms of the camel and other mammals adapted to the extremely arid lands.

The human body can lose only a small proportion of its body water before the water deficiency becomes dangerous. When, through perspiration through the skin, this loss amounts to between 6 per cent and 8 per cent of the body weight, the mouth

[1] W. J. McGee, 'Desert Thirst as a Disease', *Interstate Medical Journal* (1906), 13, 283.

becomes dry, the tongue sticks to the mouth and the senses become distorted. With a water deficit of about 10 per cent, delirium, insensitivity to pain and lack of blood at cuts and abrasions follows (cf. McGee's description). By the time the water loss has reached about 12 per cent of the body weight, recovery without medical assistance is impossible since the victim can no longer swallow. Meanwhile changes have been taking place in the circulation of the blood and reactionary changes in other organs of the body. The blood has become thicker and more viscous; the heart has to work harder to pump it round the body and slows down under the strain. As the rate of circulation is reduced the body temperature rises and death follows at a water deficit of 10–12 per cent in the desert in the summer. It is this rapid rise in body temperature following the breakdown of the body's thermostat which merits the term 'explosive heat death'. This thermostat set at approximately 37°C (98·4°F) is controlled to within about 1°F by the evaporation of sweat from the surface of the skin.

The camel, however, possesses a rather different mechanism and the thermostat has a wider range. It sweats, but to a lesser degree than the human being, and to conserve body water its temperature can rise much higher before the sweat mechanism comes into operation—to about 41°C (105·8°F). Moreover, its body cools down at night to as low as 34°C (93·2°F) so that during the heat of the day it takes longer for its temperature to rise to the critical level. One answer then is that the acceptable body temperature range of the camel, 34–41°C (93·2°–105·8°F), is greater than most warm-blooded animals—37–40°C (98·4°–104°F)—and the rapidity of onset of explosive heat death is much reduced.

The camel still, however, loses body water through sweat and through the lungs but it loses very little water through the renal tract and through the faeces. Water lost in these ways is restored from the store of the hump(s), which is almost exclusively fat and like the tail of the sheep is considered a delicacy by the indigenes of the Old World deserts. There is no water reservoir in the camel in the liquid sense and the desert traveller at the extremities of thirst would find nothing more than the objectional green digestive juices of the stomach pouches if he were to dissect his camel. The camel obtains its body moisture under conditions of water deprivation from the oxidation of the fat in the hump and other parts of its body, moisture which is lost through the lungs very soon after it has been produced.

The consumption of body fat as a moisture regulating mechanism in a camel which has no opportunity to drink or eat green fodder which contains moisture, must mean that the body weight is reduced. In experiments at Béni Abbès in the Algerian Sahara some interesting facts emerged. It was shown that a camel kept on a diet of dry dates and hay in the cool season was, at the end of sixteen days, thirsty but not in a serious condition. In June a camel weighing 450 kg (992 lb) was put on a dry diet without water for eight days; by the end of the period it was in a very distressed condition and it weighed only 350 kg (772 lb). On being allowed to drink, it consumed 103 litres (27 gallons) of water in ten minutes and regained its original body weight. It was estimated that if the camels had been worked, the limit of endurance in summer would have been about a week which far surpasses that of a human being under similar conditions. It was also shown that dromedaries grazing on fresh winter grasses showed no interest in water over a period of two months.

Obviously these figures cannot be applied as the norm. The length of time the camel can go without water will depend on the amount and moisture content of its food intake, on the work it is doing, the season of the year, the nature of the terrain and the distance it has to cover as well as the type (Bactrian or dromedary), its breed and condition. Conflicting travellers' tales thus become more comprehensible but it would appear that two humps are better than one and one is infinitely superior to none at all! It also points to the need for expertise in the conduct of camel caravans and to the great skills acquired by the camel grazing and breeding nomads who understand that the mating season of the camel coincides with the rainy season and that pregnancy lasts for twelve months, a pregnancy which must be brought to a successful term if the animals, and the people who depend on them, are to survive.

The more efficient utilisation of the Old World arid lands is thus linked to the camel with its greater load-carrying capacity than the weaker llama of the high dry plateaux of the Andes. The camel is better suited to the dry-land environment and has been used in Egypt as a domesticated animal since at least the second millennium B.C. Even so, it is not as independent of water as some of the small mammals. The jerboa can survive indefinitely on dry food alone and the American kangaroo rat even puts on weight under the same conditions.

In comparison with these animals man is poorly equipped for life in the arid lands. His anatomy and physiology require a regular water supply if the body mechanism is to continue to function; he must live beside water, carry it with him or have it brought to him if he is to utilise the desert. Compared with the camel his water intake is slow, about 1 litre (one-fifth of a gallon) at a time. He can reduce water loss by protective insulating garments, by remaining in the shade and by reducing his output of energy; it is thus more economical in the face of water shortage to travel on foot in the desert at night and lay up in the shade during the day.

Perspiration from the human body results in the accumulation of salt at the skin surface and the salty taste of sweat is well known to those who have struggled up a desert escarpment in high air temperatures or tried to excavate a vehicle from soft sand into which, by mismanagement, it has sunk up to its axles. Unless this sodium chloride is replaced, fatigue, cramp and eventually failure of the circulatory system will occur. An extra intake of salt, in tablet form for the sophisticated tourist to the hot lands, or a salt 'lick' for the indigenes, is essential and readily available in most arid lands, especially in the coastal deserts from which it is a valuable commodity of trade. Saline ground water often tends to be too purgative to provide the requisite salt intake without unpleasant side effects.

If human beings can never hope to equal the ability of the camel in endurance and water intake, evidence has accumulated which indicates that acclimatisation to the hot dry-land environment takes place over a period of time and that for the white-skinned it is a more successful acclimatisation than to the humid tropics. It has been noted that after a few days, changes occur in the structure and functioning of the sweat glands and that there are changes in pituitary and adrenal functions according to the degree of acclimatisation. Five types of progressive adaptation can be visualised: (a) decreased heat production; (b) increased ease of heat loss; (c) increased sensitivity to the heat regulatory mechanism; (d) reduction in the secondary disturbances brought about in the course of heat regulation; and (e) increased tolerance of either a rise in body temperature or of the secondary consequence of regulation.[1] It is also suspected that the native desert-dweller may possess, but to a lesser extent, the camel's ability to tolerate the disturbances of dehydration and some workers favour the

[1] D. H. K. Lee, 'Applications of Human and Animal Physiology and Ecology to Arid Zone Problems' (in) The Problems of the Arid Zone (1962), UNESCO, XVIII, p. 217.

D

view that the long slender-bodied type of human being is characteristic of the arid zone and better suited to combat heat stress.
Experiments in relation to the US Armed Forces have shown that
dark pigmentation is not necessarily an advantage in the hot arid
zone. Negroes and whites, fully clothed, walking or seated, had
roughly equal tolerance to hot dry conditions, although when
nude and exposed to the sun, sun-tanned whites had a higher
tolerance. Of course, purely environmental circumstances do not
provide the complete answer to tolerance of heat and drought in
human beings. It is the attitude of the individual or the group
which is important. The Bedouin accept the environment as their
natural home and prefer it to the oases with which they are
associated in an economic and political symbiosis; yet many
Europeans have come to love the desert (as they have the Polar
Lands) and through the ages have come to terms with it. Passarge,
Lawrence, Doughty, Philby, Glubb and Thesiger are but a few
examples (cf. F. Spenser Chapman's attitude to the hot humid
zone in *The Jungle is Neutral*).

As Montgomery wrote recently:[1] 'The main curses of life in the
desert are flies and sand and dust storms. But apart from these,
the soldiers of the Eighth Army found it, on the whole, a healthy
life—but not a soft one. Despite the heat there is a quality of the
air which gives the climate an exhilarating feeling. We who lived
and fought from Alamein to Tunisia were very fit, and in good
heart. Except for a few special ailments such as desert sores,
stomach upsets and jaundice—the latter being chiefly confined to
officers—there was little sickness.' It is probably true that the
opportunities for transmission of disease are definitely less than
in warm humid climates but malaria is certainly endemic in some
oases and problems due to the desiccation of the skin, conjunctiva
and mucous membranes are increased.

Nevertheless man cannot take liberties with aridity. The
soldiers of the desert campaigns would not have been 'fit, and in
good heart' if the engineers and the water-cart drivers had not
performed their duties efficiently to provide the half-gallon man/
day water ration and had not the soldiers learnt to economise on
water intake during the heat of the day. Half a gallon (2·3 litres)
per man per day is now considered too low for maximum efficiency and at least twice this amount is drunk by each member
of a modern caravan. Much water is undoubtedly wasted under

[1]Field Marshal Viscount Montgomery, 'The Battle of Alamein', *Sunday Times
Magazine* (24 September 1967), p. 25

hot arid conditions through the use of simple evaporative cooling mechanisms such as the 'chagul' of West Pakistan and the North-west Frontier where the water temperature is reduced by evaporation through the canvas of the container.

Without water, man is at the mercy of his environment. Burke and Wills died from thirst in Central Australia; among some Bedouin tribes it is said that men have survived only because they have drunk the contents of the stomach of a slaughtered animal. There are fifty graves at Tinajas Altas on the Camino del Diablo (Devil's Road) along the Mexican/United States border, a route used by prospective miners in the Overland Rush to California in 1849, where for 241 km (150 miles), between Sonoita and Yuma, water is available in limited quantities at only two places. (Only a very thirsty person can tolerate the squalor of a desert drinking hole!) Estimates of the deaths of travellers along this route vary from hundreds to thousands. Over 400 people are believed to have died between Sonoita and Yuma and numerous Mexicans from Sonora perished *en route* to the placer-gold discoveries in the Colorado Valley just before 1860. In the Sahara from El Gatrun to Bilma in the Kaouar there is almost complete desolation with only one well which provides poor-quality water near Bilma. Along this route came the slave-caravans from Bornu to Tripoli until the traffic was stopped by the French and the Italians. Estimates of the losses in slaves over waterless stretches such as this, amount to 80 per cent yet a profit was made out of the 20 per cent who survived, i.e. 400,000 out of an estimated 2,000,000. In 1951 a lorry convoy carrying hydrogen cylinders for meteorological balloons to be released from Kufra in the Libyan desert suffered a high loss of life. Farther to the west a lorry on its way from El Gatrun to Zouar lost its way and ran out of petrol. Two of the vehicle's crew set out on foot to find water, the third was found alive by a search party forty days later—his moisture intake had come entirely from the load of tinned fruit which the vehicle was carrying.

Yet life in the arid lands has persisted for millennia. The Bushmen of the Kalahari have learnt how to find water in apparently impossible conditions; one old woman is reported, in what is probably an apocryphal story, to have kept alive for a fortnight, herself, two men and four donkeys by sucking water from the Kalahari sands. Certainly the Bushmen have discovered the 'sip-layer' from which they suck water through hollow reeds which penetrate through the dry surface sand to the moister layer beneath. In Texas 'the ignorant traveller, certainly has to learn that there is alcohol in cactus, that if the thorns of the yucca and the prickly pear are burnt off with petrol torches, the stem of these plants possesses emergency rations for a famished cow'.[1] Yet there is no problem at Hassi Messaoud or at other oil wells nearby in the Algerian Sahara since water occurs but 9 m (30 ft) below the surface. Urbanisation of the more arid parts of the American droughtlands, mineral exploitation and irrigated agriculture all demand vast quantities of water. This water must mainly come from underground stores; there is not enough available at the surface except where the seas wash against the desert shores. Under suitable economic stimulus this sea-water can be desalinated for consumption by men, animals and industry.

[1] A. Cooke, 'The Texas Drought', *The Guardian* (7 August 1953).

5

WATER RESOURCES OF THE ARID LANDS

Clearly the most precious resource of the arid lands is water. For
vegetation, animals and man it controls their very existence, their
distribution and their density. By its almost universal absence
from the dry-land surface it clearly distinguishes the appearance
of the arid from the humid margins yet, for the utilisation of the
dry lands for pastoralism, sedentary agriculture, trade, exploitation of mineral resources and urbanisation, water must be available in dependable quantity and of the right quality for the
selected economic activity. Only two major sources are available
away from the major inland seas and the oceans. There is, firstly,
ground-water which is a dwindling resource in some of the arid
lands, and secondly the perennial rivers which receive a sufficient
supply of water from the humid zone, or from moisture islands
within the arid lands, to maintain a flow under frequently influent
conditions and a strong evaporation regime. If water is available
at the surface from springs, or can be brought to the surface by
pumping, it must be protected from evaporation where possible
until it is to be used. When ground-water is available in quantity,
and of the right quality, or when perennial streams can be tapped,
it would seem that prospects for 'making the desert bloom' are
endless and that the 'dry third' can make a substantial contribution to the world's supply of grains, fruits, meat, vegetable oils,
fibre and timber. Desalination of sea-water at the right price
might open up the coastal deserts whose proximity to the trade
arteries of the oceans, and local fishing grounds, would seem to
give them an advantage over the more inland deserts. It is, there-

fore, necessary to probe fairly deeply into the water resources of
the arid lands before looking at their utilisation.

Under the influence of gravity, rain-water percolates mainly
downward through sediments and rock, but there are two other
possible directions of movement without which there would be
far fewer emergences of ground-water in the arid lands. Water
under pressure can move upwards from deeper to higher aquifers
and can also move laterally within an aquifer on a gradient determined approximately by the dip if the aquifer is a sedimentary
rock. Not all the rain which falls at the surface will, however,
reach underground protection; in the Kalahari the addition to
ground-water is estimated to be only 4 per cent to 5 per cent of
the annual rainfall total which means that only about 1 mm
($\frac{1}{20}$ in) of rain per annum percolates from the surface. Even torrential showers at first do little to recharge ground-water reserves
since sheet run-off begins before the dry-sand surface is saturated.
The level of water in wells has been observed to remain static even
when the land surface is flooded and only then does it begin to
rise. In the drier parts of Tanganyika, where the average evapo-
transpiration rate is 85 per cent, percolation amounts to only
10 per cent while in Tunisia, of the 32·5 thousand million cubic
metres of rain-water received at the surface over the whole of the
country less than one-thirtieth enters the underground aquifers.
Apart from ascending thermal waters of volcanic origin, the upward movement of water from aquifers must be explained either
by artesian conditions as in the artesian basins of Australia and
North Africa or by local pressures in open faults or fissures. The
lateral movement along the aquifers is well demonstrated in the
oases of Egypt where the water flows with a gradient of about
1:2,000, usually but not exclusively in the Nubian sandstone, and
is under sufficient hydrostatic pressure to overflow where the
ground is dissected or has been so dislocated by tectonic movements for the overlying cap rock to be breached by the water
pressure. It is this lateral movement in the aquifers which provides the water of the extremely arid lands although, of course,
it may have to be pumped to the surface.

Although it is generally assumed that sandstone is the best
aquifer—since it has a pore-space of up to 40 per cent, observations indicate that limestone, with a pore-space of only 10 per
cent, is a better reservoir of water because of its interconnecting
joint systems; the best sandstone aquifers are also the well-jointed
sandstones. The joint systems in the limestone were considerably

enlarged during the 'pluvials', as is demonstrated in Cyrenaica by the fossil karst of the north Mediterranean coastlands, and now provide excellent interconnecting reservoirs of ground-water. Water in these limestone aquifers, which for best results should be of low density and high crystallinity, must be retained by relatively impervious rocks below and above. In North Africa and the Middle East, chalk with a high marl content acts as a retainer rock as it does for the Cretaceous Dakota sandstone of the USA where 914 m (3,000 ft) of chalk and shales overlie this most important regional aquifer. Igneous rocks, especially lava sheets, may act as retainers in parts of Australia, but in the Central Australian desert Cretaceous shales cover Jurassic and Cretaceous sandstone aquifers.

These sandstone aquifers are famous as a source of artesian water in the USA, southern Africa and Australia. Through them the flow of water is slower than the sometimes turbulent flow in the fossil karst caverns and tunnels—but the yield is high, as much as 800–1,000 m^3 per hour. Aquifers are represented at some depth in Saharan Africa by the early Mesozoic Nubian sandstone which feeds oases such as Dakhla in the Egyptian desert. Here clay overlies the aquifer and must be penetrated to depths of 30–91 m (100–300 ft), according to locality, before the wells reach the water table. At Farafra, north-west of Dakhla, the Nubian sandstone dips more slowly than the gradient of the water table and the water overflows into limestones and marls. In North America the quite thin—91 m (up to 300 ft)—Dakota sandstone of Cretaceous age corresponds, as a main aquifer, to the Jurassic/Cretaceous series of Eastern Australia.

The major artesian basins, in which the sandstone aquifers provide the ground-water store, consist of sedimentary basins on a semi-continental scale occupying down-sags in the continental basements and shields of Gondwanaland. They are margined by flexures or upwarps which accentuated the dip of the sedimentary aquifers and retainer rocks—typical are the cuesta structures of the margins of the Saharan massifs where the ground-water emerges in sub-artesian fashion at springs on the margin of the escarpment and gives rise to water-holes, or in some cases to small oases. Dakhla oasis belongs to this category together with Kharga farther to the east. In Australia, the artesian basins cover more than one-third of the continent. The Great Australian Basin has a latitudinal extent of over 1,930 km (1,200 miles) while from west to east it covers about 1,448 km (900 miles). To the east the catchment area of the aquifer is along the Great Divide, while to the west it appears at the surface in Lake Eyre and forms crater-like mounds elsewhere near Coward Springs. This Great Artesian Basin covers about 1,554,000 sq km (600,000 sq miles) and its presence is critical in the utilisation of the arid lands of Queens-land, New South Wales and South Australia.

In comparison, the crystalline shields of the continents, which fortunately outcrop to a much lesser extent than their sedimentary cover, offer only local rather than regional resources of ground-water. Here the water is found only in well-jointed rocks or in rocks which, in the arid zone, still carry a fossil chemically de-composed regolith. The yield is generally low (1–4 m^3) per hour; there is often no continuous water table and supply can be ob-tained only from individual wells which tap reserves in the joint systems. Only rarely does this ground-water emerge as springs. This is the typical water-supply situation of areas like the Sudan, the dry parts of East Africa and in parts of India where shallow wells, excavated with difficulty in the solid rock offer, without doubt, the poorest prospects of ground-water supply in the arid lands. Volcanic terrains provide rather better possibilities but the quantity of ground-water available depends on the nature of the lava sheets and the discovery of supplies is often a matter of luck rather than judgement. Best results occur when the lavas are well-jointed and vesicular, when fault shattering is frequent, when intercalated tuffs provide aquifers and when the permeable lava sheets have a good retaining stratum of low permeability sedi-mentary rocks or less permeable sheets of lava. Under the best conditions, where a faulted lava escarpment overlies a sedimen-tary basement, good yields may be obtained, as in the Yemen, but complications are often introduced when dykes divide the lavas into compartments some of which are dry while those adjacent are rich in water.

The use of regional sedimentary aquifers or of the local occur-rences from crystalline igneous rocks depended, in earlier times, very largely on favourable structural arrangements, as in the artesian or sub-artesian situations, or where the degree of dissec-tion or faulting produced a topographic situation in which the ground surface encountered the water table. The emergence of ground-water was localised and its utilisation expressed by the discrete and often widely spaced water-holes and oases. Some-times the oases are arranged in a linear pattern, as where water emerges at the base of a fault or fault-line or lava escarpment, or along a fault-line without associated topographic break, but often the arrangement is haphazard and vegetation, pastoral and agricultural utilisation and population distribution lacks continuity. Without elaborate bore-hole equipment to penetrate the retaining rocks to the aquifers beneath, the less technically endowed civilisations and the pioneer European agriculturalists of the arid lands naturally exploited those zones which, through the indices of vegetation, offered more reliable prospects of ground-water. These were the zones of unconsolidated sediments which can be penetrated with relatively simple hand tools; they coincide broadly with the forms of accumulation outlined in Chapter 3, i.e. the detritus-filled intermontane basins, the wadi floors, the piedmont coalescing alluvial fans or *bajadas*, flood plains and deltas. These are the late-Tertiary and Quaternary gravels, silts, clays and partly consolidated sandstones which are generally highly permeable but for which the solid rocks of the basement or lenses of impermeable sediments offer water traps. Normally the yield from such reservoirs is not as high as from artesian and sub-artesian aquifers but there are two situations in which yields and reserves may be increased.

In an aggraded landscape, such as the High Plains of North America, where alluviation has occurred on a regional scale, there are numerous examples of the interment of rock domes and ridges beneath unconsolidated sediments. These act as traps to the movement of ground-water and may even provide artesian or sub-artesian conditions. They are geographically and perhaps genetically related to the semi-arid or steppe lands rather than to the extreme deserts. The High Pampas of Argentina and the major tectonic units of the Ganges/Indus basins and the lowlands of the Tigris and Euphrates in Mesopotamia contain examples of this type. In volcanic areas the ground-water traps in unconsolidated sediments are often created by individual dykes behind which water can accumulate and be tapped over a wide area. The best dyke traps are found where the volcanic material was much more resistant than the surrounding country rock and, in consequence, stood well above the surrounding landscape before the phase of alluviation began. The more regular the top of the dyke the more effective a trap it becomes. Unfortunately dykes do not often occur singly; they may radiate from a volcanic neck or occur in 'swarms' and, under these conditions, they divide the ground-water supplies into separate hydrological reservoirs which give different yields varying from the very good to the very poor. There are many examples of this type in southern Africa.

D*

The detritus-filled basins of the Old and New World Tertiary
mountain systems offer ideal conditions for ground-water
accumulation especially where structural traps are available. The
Basin-Range province of the USA consists of fault blocks and
fault-angle or rift valley depressions filled with sand and gravel
over-lying 'caliche' which acts as a retainer and, incidentally,
reduces the volume of sediments available for water storage to the
layers above the 'caliche' except where it has been breached by the
down-cutting of streams from the mountains to reach the axial river
or playa depression. Further complications arise from the stratigraphy of the sediments which may be uniformly bedded with a
gradation of facies from coarse gravels on the margins through
sands to clays in the centre of the depression; frequently, however, there is much interfingering of permeable and impermeable
materials with 'perched' water tables and complicated inter-
connections between aquifers. The level of water in wells thus
varies widely within the same locality, and towards the centres of
these depressions there have been many unproductive bores
expensively sunk with little or no ground-water return. The recharge of the aquifers comes not from rainfall within the depression but from the mountains on the margins and it is from the
coarser material of the piedmont alluvial fans that the most
sustained yields are derived.

Somewhat comparable conditions occur, for instance, in the
Tsaidam Basin on the margin of the hot and cold deserts south-
east of Lop Nor in the Takla Makan of Central Asia. This is a
tectonic depression of synclinal origin aggraded by Tertiary and
Quaternary deposits so that it now has the appearance of a vast
plain with occasional islands of unburied hills rising above the
fill especially in the higher north-western section 2,800–3,000 m
(9,000–10,000 ft) above sea-level. The lower central section,
2,600–2,800 m (8,500–9,000 ft) in altitude, is almost perfectly
plane and made up of alluvium (of high alkali content) and
mobile sands of lacustrine and deltaic origin. To the south and
east of this central plain there is a different and more complicated
landscape produced by contemporary streams draining from the
eastern Kouen-Lun; this complexity of relief is repeated in the
complex stratigraphy of the sediments and increases the difficulty
of finding underground water supplies compared with the higher
sections to the north and west.

The sediments of coastal plains may be important sources of
ground-water but as in the 'bolsons', the disposition of the

aquifers is often extremely complicated. The coastal plains of arid
lands were much affected by the Quaternary oscillations of sea-
level so that marine and continental facies are found in close
proximity with aquifers at varying levels and of differing yields.
The stratigraphy of these coastlands requires careful investigation, in which electrical resistivity methods have proved valuable,
and accurate interpretation if sufficient ground-water supplies are
to be tapped in an economic way. Good results have, however,
been obtained as in the Philistanean plains of Israel where yields
of up to 100 m³ per hour provide water for irrigation purposes.

Before intensive utilisation of ground-water is possible, it is
essential to determine not only that ground-water is available but
also that it is capable of providing a sustained yield. No longer is
it simply a question of finding water to supply the needs of one
household; the nature of the problem is now 'how much' and for
'how long' at the rate of withdrawal of the particular economic
activity. In the humid zone where the ground-water is constantly
recharged from rainfall there is normally no problem except
where modern water-using industries have exceeded the rate of
recharge of the aquifers from which they are pumping supplies.
If pumping were to cease then the ground-water reservoirs would
be recharged. In the arid zone, however, rainfall is so slight and
evaporation rates so high that, except under exceptional conditions, there is little increment to the ground-water capital and
it is fast becoming a dwindling asset. Many users of ground-water
in the arid zone are, in fact, living on the capital provided by the
'pluvials' of Pleistocene times and the water which is being used
for irrigation, for stock rearing or for human consumption is
'fossil' water. Over-utilisation of ground-water resources is expressed in the declining yield of wells and the lowering of the
water table through pumping to the stage when it becomes uneconomic to sink deeper bore-holes. This over-utilisation is often
the result of the lack of understanding of the ground-water
situation partly because of the lack of dissemination of knowledge
which already exists and partly because the use of ground-water
has been based on uncontrolled economic activity according to
need and technical resources. The two great necessities in the arid
zone are therefore the assessment of the ground-water budget,
which, because of the wide areal extent of the aquifers, may
require international cooperation, and the recharging of ground-
water supplies. For an inventory of ground-water resources, the
losses must be determined by maintaining records of the amount

of water abstracted. For instance, in the nine basins of the Gila River and its tributaries in Arizona, the annual rate of withdrawal increased year by year from some 1,110,132,000 cu m (900,000 acre-feet) (one acre-foot or 1,233·48 m³ equals the volume of water necessary to flood an acre of level ground to a depth of one foot, equals 272,000 gallons) to 3,919,999,440 cu m (3,178,000 acre-feet) in 1950; this rate of pumping was estimated to be at least thirty times the annual rate of replenishment. The problems in assessing the amount of withdrawal in the oases of the Sahara, Libya and Arabia are obviously immense but a ground-water resources budget is essential before development can take place; it is probable that for many areas more is known about the rate of recharge than the rate of withdrawal.

Now that the problem of recharge has been formulated various techniques have been developed to discover and utilise the best topographic, stratigraphic and hydrological situations. Floods must be diverted to the areas of most rapid infiltration or prevented from travelling so far on the surface that evaporation extracts more water than is necessary. For this purpose simple dams have often been constructed near the apices of alluvial fans to encourage the water to infiltrate rapidly through the coarser deposits near the hill foot. In the USA detention reservoirs have been constructed to hold back floodwaters for quick percolation as in Los Angeles County in California.

Fortunately some of the major aquifers on which future utilisation of large stretches of the arid lands will depend are charged in the seasonal rainfall margins of the arid lands. In Algeria the artesian system, on which exploitation of the Algerian oil fields relies, is constantly replenished by the rains and snow of the Atlas; in the Tsaidam Basin in Central Asia melting snow from the surrounding mountains saturates the sands and gravels and is betrayed by the more luxuriant vegetation. When natural recharge is available the problem is often to ensure that water is not wasted by unutilised seepage. In this context the *schotts* of Algeria and Tunisia are good examples. They represent the great evaporating machines of the Turonian aquifers from which many thousands of millions of gallons of water are lost each year and in which huge deposits of salts accumulate.

Salinity problems

The water which percolates underground inevitably takes into solution minerals from the rocks and sediments through which it passes. When it emerges at the surface as springs, is drawn up from wells or pumped from bore-holes it is often highly mineralised. For vegetation, animals and man there are limits to the amount of minerals which can be tolerated. For human consumption 'sweet water', as it is described in the Egyptian and Libyan deserts, is normally required in contrast to the brackish or saline water containing a large proportion of sodium chloride and other dissolved salts. In an arid climate water with a salinity of 3,000 parts per million of sodium chloride can be drunk regularly without ill effects but when the proportion rises to 5,000 it can be tolerated for only short periods. If salts other than sodium chloride are present then the allowable proportion is much less. In Jalo, an oasis in the northern Libyan desert, for example, the water has a pH of 7·4 with a high permanent hardness of 130·2 (i.e. the hardness cannot be removed by boiling). The salinity is 3,880 parts per million which is higher than is medically advisable especially when magnesium sulphate, calcium sulphate and calcium chloride are present in significant quantities as the following table indicates:

DISSOLVED SALTS IN WATER AT JALO OASIS

Parts per million

Calcium	336
Magnesium	132
Potassium	23
Sodium	834
Chloride	1,560
Sulphate	907

The wealthy traveller can have sweet water brought from Botafal about 40 km (25 miles) away on the track to Kufra, but the inhabitants must either drink oasis water which is sometimes organically polluted or preferably bring in water three-quarters of a mile from wells which have not been polluted by animals. In a regionally saline zone, good drinking water with salinities of 200–2,000 parts per million may be found after rain-showers since the newly percolated water does not mix readily with the saline water beneath, but the yield is small and the distribution very sporadic even over distances of one hundred yards.

For domestic animals the sodium chloride content can be considerably higher as long as there is no magnesium sulphate present. In southern Australia horses thrive on water with 6,260 parts per million and upper limits are stated to be up to 7,800 parts per million. Sheep have the highest salt tolerance, 15,600 per million, and cattle about 9,400 per million. Waters with a much higher salt content than the maxima men and animals can withstand are widespread in the arid lands and reach very high levels in the inland seas of Asia and, of course, in waters which margin the coastal deserts. Desalination of ground- and sea-water is thus a most important technical problem which must be solved economically if stock rearing and high population densities, such as result from urbanisation, are to be increased.

The problem of salt tolerance in cultivated crops is very complicated since saline ground-waters and salt-affected soils are intimately related. The control of ground-water, or more specifically the control of the level of the ground-water table, is the basis of irrigation farming and the reclamation of salt-affected ground. Inadequate drainage has been held responsible for the sterilisation of land by salt in many arid lands from the USA to India; only the most open-textured soils, which do not retain adequate soil moisture for plant growth, are immune. It was shown earlier that some plants of the arid zone have mechanisms to resist the toxic effects of saline soils and ground-water; in contrast, the cultivated crops of the arid lands are rather sensitive to salinity and a relatively low figure of 700 parts per million has been accepted as the norm of water suitable for irrigation. This compares very unfavourably with the tolerance shown by men and stock for whom more brackish water supplies are suitable. It means that mechanical or chemical desalination of water for irrigation purposes is completely uneconomical and that amelioration of salinity problems for crops must be tackled along entirely different lines by controlled irrigation and flushing methods or by selective breeding of cultivated crops with a high degree of salt tolerance. According to some investigators, plants may acquire a degree of salt tolerance which can be transmitted to their progeny if the seed from adapted individuals is used. Other attempts have involved soaking seed in salt solution for several hours before sowing which produced increased yields from cotton on saline soils in the USSR but did not demonstrate increased salt tolerance; in fact, when applied to wheat and barley in Pakistan this treatment was not effective. Some success has been obtained by increasing the soil moisture for sugar beet and barley when the crops have shown greater tolerance to salinity. In general terms, vegetables are moderately tolerant to salt with asparagus and spinach at the most tolerant end of the scale while radish, celery and green beans are sensitive. Of the field crops rice is tolerant and is an important crop for saline soils—the special moisture conditions necessary for its growth, i.e. a flooded soil, help by reducing the salt concentration in the 'wet zone'. Barley, sugar-beets and cotton are also grown on saline land and some grasses have proved to be salt-tolerant species. Apart from the date palm, and to a lesser extent the vine, most fruit trees are sensitive to saline soils and ground-water. Fruit trees, for instance, do not thrive on the solonetz soils of the Ukraine, although birch, willow and aspen may grow on such soils elsewhere in the USSR. A short summary such as this can do no more than indicate some of the problems faced by the cultivator in the arid lands through saline ground-water and salt-affected soils. It does, however, point to the need for the research into salt tolerance of economically important crops and fruit trees, research which is in hand in many parts of the world.

Desalination is, of course, possible and has already reached a stage in technological development where, for drinking water, the price per litre/gallon is not very much higher than that obtained through a water-board undertaking in the humid lands. The salt-water conversion plant in Guernsey in the Channel Islands is an example in the humid zone while other islands with small catchment areas such as Curaçao obtain fresh water from sea-water. All desalination involves high capital expenditure, large amounts of fuel or other energy and a skilled, if small, labour force. Other than in the oil-rich countries of the dry Middle East, these are scarce with the exception of the energy available from insolation. Desalination of water may be achieved through distillation, by freezing or chemical absorption since water has great stability, i.e. exposure to extremes of temperature or to chemicals does not result in permanent modification of its chemical composition. Since it has a high electrical resistivity, ionised mineral salts may be removed by passing an electric current through the water, i.e. by electrodialysis. Distillation is the natural method of desalination accomplished through the effect of insolation on the oceans through the medium of the normal atmospheric cycle. To reproduce this process artificially requires expensive fuels unless solar energy is recruited as the source of heat. The Russians have employed, in the Kara Kum, cylindrical mirrors to focus the sun's rays on a glass tube which acts as a boiler from which 75,000 tons of distilled water per annum have been used for livestock. Elsewhere attention has been directed mainly to multiple flash-stage distillation, as in Guernsey, where several evaporator-condenser units are employed in series. A higher yield of distilled water per unit of fuel consumed can be achieved through vapour compression operating on the heat-pump principle but these have been manufactured only in small units, driven by diesel engines. They were used in the North African campaign in the Second World War and are currently employed in desert oil-field exploitation. Freezing separation systems depend on the fact that the dissolved salts do not solidify when ice is formed in saline water at 0°C. When the ice is removed from the brine it will melt into pure water. Here refrigeration processes are involved together with mechanical devices to separate the ice from the brine. Freeze-separation has been employed in the cold winter deserts of the USSR during the frost period, by flooding saline water into artificial ponds and draining off the brine when the ice thickness is about half an inch. In this method the yield varies according to the degree of frost—40–50 litres per square metre per day at —5°C to 120–160 litres per square metre per day at —20°C. To be of value suitable storage arrangements must be provided to conserve the water for the hot dry months. Other methods are being investigated and used at the pilot plant stage. They include the reversed osmosis system in which water under pressure is forced through membranes which filter out the unwanted ions; ion exchange by the addition of chemicals and osmionic separation are also being further investigated. It is clear that all these methods are unlikely to produce the relatively pure water required for crop irrigation on a large scale at a cost which could be met by the subsistence farmer but they will be used increasingly for urban populations, probably with nuclear power as the main energy source, in the future. A very large sea-water distillation plant to produce five million gallons per day has been installed at Kuwait using natural gas as the fuel from the oil-fields twenty-five miles away. Even with this cheap, and otherwise wasted, source of power, the salt-free water is still too expensive for the commercial irrigation of crops. It does, however, stand in great contrast to the town of Moharek on the Persian Gulf which, at the end of the nineteenth century, obtained its fresh-water supply from undersea springs. At high water, when there was a fathom of water over the spring, divers filled skins through bamboo tubes while at low water the women were able to fill the skins directly.

In comparison, those sections of the arid lands watered with perennial streams are extremely fortunate. The Tigris–Euphrates, the Nile and the Indus have provided sufficient water over the whole of the period of human history to maintain 'the hydraulic civilisations' of the Old World, while in the New World the Colorado River offered similar opportunities which have more recently been intensively exploited. The annual water supply from the mainstream of the Tigris alone is 17,026 million cubic metres; if to this is added the waters of the Greater and Lesser Zab, Ahdaim, the Diyalah and, of course, the Euphrates, then the total annual supply of the Tigris–Euphrates Basin is just over 69,000 million cubic metres per annum. There have, however, been occasions in which areas in the basins had insufficient irrigation water to meet agricultural requirements and it must be remembered that in Iraq alone the cattle consume over 125 million cubic metres of water per annum, most of which is derived from irrigation ditches. Every year 80,000 million cubic metres of water enter Egypt from the Nile yet the demands are such that, with an increase in land under cultivation and increase in irrigation, it has been necessary to build major storage reservoirs during the last hundred years.

Utilisation of the waters of perennial streams depends to a large extent on their regimes, which is controlled by the meteorological conditions of the headwaters rather than those of the lower reaches where the losses, through evaporation or infiltration or through the abstraction of water for irrigation purposes, mainly occur. In the Nile, fed by tropical summer rainfall from the mountains of Abyssinia and from the reservoir of Lake Victoria and Lake Albert in the East African Highlands, there is a wide variation between the discharge according to season. During the floods water is available in excess of irrigation requirements which through the centuries went to waste by evaporation in flooded bottom lands and in the sea. In contrast, the early summer discharge is too low for the requirements of cultivation. It is fortunate that the Blue and the White Nile are complementary in their regimes; in August the Blue Nile's discharge is about three and a half times greater than that of the White Nile, yet in May the White Nile has a discharge five times greater than the Blue Nile and for long provided the water for the summer crops in Egypt. Without the building of barrages and dams the development of perennial irrigation would have been impossible. Similarly it was necessary to construct dams and storage reservoirs on the Gila–Colorado system to even out the regimes with the result that for

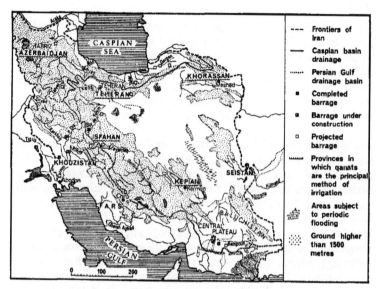

Fig. 9. Aspects of irrigation in Iran. (After Bémont, *Annls. Géogr.*, 1961, *70*, pp. 597–620)

many years no water has entered the Colorado from the Gila at the junction near Yuma in Arizona since the water has been trapped upstream at dams such as the Coolidge and Gillespie on the Gila, and the Roosevelt on the Salt River tributary to the Gila. In Central Asia and Transcaucasia the rivers flow at high-water stage in the spring from melting snow and in the spring and summer from melting glaciers. Here a redistribution of the water stocks available from different rivers at different times of the year has led to the construction of dams and interlink canals, such as the Great Fergansky which carries water from the Narin River to the less well endowed streams of the Fergansky valley in the Uzbek Republic. This project is comparable with the major barrages and canal schemes of the Indus and its tributaries (fig. 10), whose regimes are rather similar and which began with the barrage across the Ravi in 1859 and the construction of the Upper Bari Doab canal. The regimes of the Tigris and Euphrates are nourished by the spring and summer snow melt of the mountains in the north and here the main floods occur during summer.

Attempts to regularise the availability of water from perennial streams, to conserve rain-water in reservoirs and cisterns or water pumped into tanks from ground-water for stock rearing involve problems of siltation of reservoirs and reduction of evaporation and seepage. The cisterns hewn into the non-porous Eocene limestone rocks of the Jebel el Akhdar in northern Cyrenaica as at Saf-Saf (fig. 2) near Cyrene were covered by a roof of limestone blocks shaped to a standard size in nearby quarries. Evaporation from these and more modern reservoirs must be reduced if water is not to be wasted—evaporation may account for up to 50 per cent of the water loss in shallow reservoirs and up to 20 per cent in deep reservoirs. It has been shown in South Australia that a storage ratio of less than 6 to 1 is uneconomic (storage ratio =

$$\frac{\text{Volume of Water} + \text{Volume of Reservoir}}{\text{Volume of Reservoir}} : 1.$$ The modern re-

placement for a masonry cover, impossible to achieve for large reservoirs, is to cover the water with a thin chemical film of Hexadecanol (Cetyl alcohol), which will cover the reservoir if spread as a thin trail from a boat. Unfortunately this technique seems to be restricted to relatively small reservoirs (1 hectare or 2·47 acres). Only a small fetch is permissible otherwise the waves become too large to maintain the continuity of the film. Attempts to counter seepage have included the laying of thin plastic sheets on the floors of small reservoirs with a layer of earth to prevent damage by the hooves of stock, and the sealing of the pore-spaces in the soil and regolith with chemicals and by the application of clays and soil cement. One must expect further advances in this field of reducing evaporation and seepage.

This has been merely an outline of the problems of water resources in the dry lands. Water is available below the surface from perennial and ephemeral streams and unconcentrated run-off from ephemeral showers. Utilisation of these resources for a wide range of economic activities according to economic and social circumstances has been taking place since the beginnings of civilisation. What kinds of land-use are possible in the arid lands? Traditionally it was related to stock rearing, often nomadic, and subsistence irrigation agriculture with a marked difference between the ways of life and standard of living. Now the deserts are becoming urbanised and the seats of commercial agriculture and recreation while the nomadic way of life is being modified or has completely disappeared from its traditional habitats. In each case the scale of development is related to the availability and conservation of water—it is this preoccupation with water resources that helps to differentiate the arid from the humid zone.

6

CULTIVATION IN THE DRY LANDS

By many people, cultivation in the arid lands is thought to be restricted to the oases and riverine lands of the Old World or to the result of the modern, highly capitalised and elaborate systems of irrigation which stemmed from the American development of the dry south-west, the British irrigation engineers' projects in the north-west of India, the works of the French in North Africa or the mid-twentieth-century developments on a massive scale on the Nile in Egypt and the Sudan. While it is undoubtedly true that irrigation is a basic technological response to the environmental conditions of the arid lands and stretches back into prehistory, it requires a high degree of social, technical and legal organisation for success. However, the cultivation of crops without irrigation in areas of low soil and atmospheric humidity is also possible and is of equally long standing. It might be argued that such farming without irrigation demands an even more intimate knowledge of environment and the niceties of maintaining a balance between the conservation of soil moisture and crop-water demands. Yet both types of crop production have maintained comparatively large numbers of people at both subsistence and commercial farming levels and both have been attended by crop failure, economic disasters and famines. Similarly both techniques have resulted in major transformations of the environment, the one by destruction of the soil resources as when humid farming techniques were applied indiscriminately, especially on the fluctuating arid-humid margin, and the other by sterilisation of the soils by waterlogging and salination. Numerous examples drawn from both the Old and the New Worlds indicate that man's relations with his dry-land environment have been fraught with the difficulties which might be expected, given the climatic hazards. They also indicate that his actions have accentuated the problems and sometimes the intensities of aridity.

It is therefore somewhat paradoxical that the origins of agricultural activity, both in cultivation and stock farming, should have been established, but not without much controversy, in the dry lands of south-west Asia and in parallel areas in the New World. In these areas, especially in south-west Asia, archaeologists have discovered the earliest, and relatively undisputed, evidence for the domestication of cereals—wheat and barley, as also of animals—sheep, pigs, goats and cattle. It is difficult to establish with certainty the origins of cereal cultivation although it is possible to distinguish between wild and cultivated grains since, for instance, the cultivated wheat seed is smaller and less coarse than that in the wild state. It is seldom, however, that seeds are preserved in archaeological sites unless embedded in pottery, or in a charred form, although the discovery of flint sickles and grinding stones indicates that grain was used as food, albeit perhaps from the wild state. It is probable that both hunters and domesticators of animals and plants collected seeds from the seasonally abundant wild wheat and barley. Since wild grasses have dispersal mechanisms arranged for the rapid distribution of the seeds to ensure wide propagation and survival of the species, early man is most likely to have harvested those which, for genetic reasons, retained their seeds longest on the spike. Continuation of this system of selection, when the idea of sowing the surplus from the preceding harvest was established, would have reduced the number of quick dispersal types in his patches of cultivation.

In his appraisal of environment he would also have been aware that certain combinations of soils and moisture favoured growth and yields, and that these coincided with areas which offered better settlement sites, including the availability of water for himself and his animals, than those zones at higher altitudes where rainfall was more reliable and where the wild grains had been discovered in their natural state. The transference of the grasses to these new habitats is presumed by some botanists to have produced further changes in the structure of the grain-yielding plants by mutations and hybridisation to the extent that the cultivated variety could not tolerate an environment such as that from which its wild predecessors had been drawn. By about the fifth millennium B.C., these grains were being grown on the alluvial lowlands of Mesopotamia as well as on the alluvial fans and detritus spreads of hill margins and mountain basins.

One can also discern in this area the development in techniques of cultivation in the dry lands from farming based on the seasonal occurrence of winter and spring rains in the hills—the amount and reliability of which was, and still is, beyond the limits of technical control, i.e. 'rain cultivation' or dry farming to the flood-channel farming which utilises irregular run-off to reinforce the sporadic rainfall. Finally, in the vicinity of perennial streams or points of emergence of copious supplies of ground-water, it was possible to utilise areas on a permanent basis, subject to the hazards of major floods, through the arrangement and maintenance of a system of channels to convey water when and where it was required, i.e. by irrigation farming in the true sense of the term. It is this framework of techniques for the production of cereals, fruits and vegetables that allows us to establish a satisfactory system of classification and description of cultivation in the arid lands.

The antiquity of *rain farming*, or *dry farming* as it is now more usually called, is well established, but the techniques have undergone various stages of evolution in both the Old and the New Worlds, including Australia, through trial and error procedures all aimed at producing the maximum amount of dry vegetable matter in the shortest possible growing season, with the minimum quantity of water derived from rain-showers or of conserved soil moisture. Wheat, barley, maize, millets and sorghums are the best known of these dry-farming crops and are more important than the leguminous crops, but lucerne, alfalfa, clover and peas have also been successfully cultivated in various places. Nevertheless, whatever the techniques and whatever the crops, dry farming cannot be successful in the truly arid lands and the bulk of effective dry farming takes place in the semi-arid lands and more especially on the arid/humid margin, a zone in which have also occurred the greatest failures in land-use.

The principles of moisture conservation are basic to dry-farming practices. They are designed to ensure that the precipitation sinks quickly into the soil to prevent evaporation and to be held there until it is required by the crop. Water losses through transpiration by weeds must be avoided by keeping the ground clean even before the seed is sown and the sowing must take place as soon as possible in order that drought-evading crops can make use of the available soil moisture. Since supplies of soil moisture, even with the best cultivation techniques, are low the crops must be sown either thinly or widely spaced. The sandier soils with a higher percentage of pore-spaces offer many advantages both for the rapid penetration of water from rain showers and for the reduction of evaporation after the shower has passed, since the moisture must pass as a vapour through the open pore-spaces near the surface rather than by capillarity. Where the soil is less sandy in texture it becomes even more important to maintain an air-dry, weed-free surface layer which accounts for the frequent scratching of the surface by the shallow ploughs of the subsistence wheat and barley cultivators of the semi-arid Mediterranean lands of the Old World. The effect is to produce a dry mulch which may be augmented not only by leaving the stones in place in the fields but also by spreading stones or straw on the ground surface which also reduces wind blow. Where, however, the soils have a high percentage of fines, too much shallow ploughing may create a dust mulch which, under the impact of the rain-drops of the next showers, will be packed down into the pore-spaces, create an almost impermeable surface layer and reduce the amount of penetration of surface water. During periods of drought the fines are picked up by the wind as clouds of dust and blown to leeward to sterilise large areas. Clearly a nice balance has to be struck between the depth and frequency of ploughing and that this balance has not always been achieved, is demonstrated by the abandonment of land, as in the USA, either by soil erosion or because of inadequate economic returns. On the other hand, certain highly specialised forms of dry farming have been success-ful in apparently unpromising environmental situations as in the cultivation of maize and beans by the Hopi Indians of the south-west USA on sand-dunes, already noted as being a valuable source of short grasses for grazing by the animals of the nomadic peoples of the Sahara.

Experiments at numerous agricultural stations in the dry lands have demonstrated that, at the time of the harvest, the soil moisture has been completely exhausted by crop demands so that there is no reserve to be carried over to the next growing season. To carry moisture over from one year to the next is an obvious requirement if good yields are to be obtained from the favourable location of plant nutrients in the upper layers of the soil, a measure which was discovered early by the dry-farming cultivators of the Old World, through the system of fallowing.

Fallowing, one year in two or three, is an essential component of grain cultivation, that is the sacrificing of one year's crop on a piece of land to ensure a good crop the following year, a practice which in turn leads to a form of shifting cultivation for subsistence cultivators and which was ignored, with sad results, by the early European farmers of the American West. Fallowing is, however, not a passive process if the best results are to be achieved but requires constant husbandry. After the harvest, the stubble is left in the fields and the problem arises when the weeds which would transpire the carefully conserved soil moisture begin to grow. Ploughing keeps the ground free from weeds and prevents the compaction of the surface into an impermeable layer, but it exposes the ground surface to wind erosion and rain-drop impact. The alternative adopted, where suitable machines are available, is the ploughless fallow in which the weeds are kept down by machines which destroy the weeds without interfering with the protective stubble. Fallowing seems also to result in an increase in the nitrogen in the soil as a result of the activity of bacteria in the favourable soil moisture and temperature of the fallowed land. Such fallowing accentuates the patchiness of the cultivated landscapes of the semi-arid lands, especially in the Old World, and is often accompanied by alternative exploitation either by stock rearing or, when forest resources are available, as in parts of the Mediterranean Basin, by intensive utilisation of forest products such as mast in beech forests for the feeding of swine, or the cutting of cork and the gathering of figs.

From fallowing the next stage in the improvement of dry-farming techniques is the introduction of crop rotation with the fallow. Such rotations may extend over periods of several years and usually include a grain crop, a widely spaced hoed crop, forage crops and legumes as well as fallow. In the Great Plains of the USA maize and wheat have proved to be a good combination in the first two years of the rotation since the maize grown in rows conserves moisture in the bare land between. In Australia wheat is the most important crop; it may be followed by oats and fallow, or oats and a green fodder crop, the latter indicating the importance of mixed-farming techniques in some dry-land areas.

Even under the most favourable conditions, dry farming is necessarily an extensive type of cultivation which in general does not warrant, or permit for reasons of cost, the application of manures and artificial fertilisers since the amount of available soil moisture cannot sustain the increased vegetative growth. Yields are low and to produce harvests comparable with the humid lands much larger areas must be cropped. Nowhere is this better shown than in the history of the utilisation of the semi-arid lands of the American south-west where the earliest land grants under the Homestead Acts provided quite inadequate areas for cultivation. The law of 1862 provided 64·75 hectares (160 acres) for each settler after five years' residence and use, established perhaps under particularly favourable rainfall cycles which gave short-lived optimism concerning the successful exploitation of the area by dry-farming techniques in a zone which was then mainly used by open-range graziers based on perennial streams or good springs. The supremacy of the rancher which reached its zenith in 1882 declined sharply as the cattle boom subsided and the home-steaders flocked into the comparatively waterless grasslands—into a zone which had been investigated by a Federal Commission headed by J. W. Powell. The conclusions of this Commission showed that, while 160 acres was viable for subsistence or commercial agriculture for a family in the humid east, it was totally inadequate for the dry lands of the west. Powell recommended not less than 10 sq km (4 sq miles) (i.e. 2,560 acres) as a minimum for each homestead and that the terrain should be scientifically appraised before settlement took place. He emphasised that land should not be allocated in rectangular plots but should take into account the pattern of drainage basins. The failure of the American dry-farming settlement must be attributed to the failure of the government to implement these recommendations and to the persuasive writings of propagandists who encouraged home-steaders to farm land with completely unsuitable techniques of humid land farming. It proved also just how poor was the perception of environmental hazards and how easy it was to destroy the basic environmental resources of soil and vegetation. In comparison the more integrated dry-farming systems of North Africa have survived successfully to the present day and account for the greater part of the land under cultivation; maize is grown without irrigation in the Maghrib on the Atlantic coast of Morocco but probably the most remarkable example of dry farming has been demonstrated at Sfax in Tunisia where extensive olive groves have been developed on the favourable sandy soils with rainfalls of less than 254 mm (10 in) per annum. It is interesting that Powell was also of the opinion that successful dry-land farming is possible only on areas of sandy soils.

Reinforcement of the soil moisture from showers by the utilisation of concentrated or unconcentrated run-off, leads to

consideration of the farming techniques which cover a wide
spectrum of water-using devices linked intimately with the details
of the terrain, river regimes, levels of technology, social organisation and financial resources. At one end of the scale are the
primitive dams thrown across a diminutive gully in which water
may flow one year in ten; at the other are to be found the huge
dams built on the world's major perennial rivers such as the Nile,
Indus and Colorado. The level at which run-off assisted dry
farming ends and irrigation farming begins is difficult to define
except in terms of the scale of human activity involved. When the
water is supplied to the 'fields' by floodwaters from seasonal
showers on slopes, alluvial fans or in seasonally filled river
channels, then the farming is probably a variant of dry farming
under special conditions of terrain even though it may involve
some primitive attempts at water control and diversion. Such
farming is far removed from true irrigation farming and is often
described as *flood farming*, or better, *flood-channel* farming.

The simplest techniques are found in many parts of the Old
and New Worlds in, for instance, Tunisia or Tripolitania, or the
mountains of Iran and Baluchistan, or in earliest times in south-
west Colorado. Across the beds of tiny ephemeral streams, or even
shallow depressions of the hill slopes, are erected check dams
built of stones and brushwood whose effect is to hold back both
concentrated and sheet run-off and to reduce the intensity of
linear erosion and the rate of areal erosion. The silt which would
otherwise be washed down the slopes and channels is retained
behind these dams and provides a suitable medium in which crops
can be grown. Seasonal or sporadic floods build up new layers of
silt to restore fertility while the problem of exceptionally heavy
showers breaching these primitive dams was solved by arranging
dams in tandem down the slopes. Such a technique involves no
more than retaining moisture for a longer period than would
otherwise occur but may be elaborated into a system of hillside
terracing arranged to reduce the velocity of run-off down the
slopes and to increase the time available for penetration.

The larger river beds in areas of seasonal rainfall offer a less
precarious flood-farming system in terms of reliability of water
supply but pose problems of the control of destructive torrents in
that the deposits of silt among the larger gravels and boulders may
be covered or swept away by the next floods. Nevertheless, successful flood-channel (rather than flood-plain) farming has been
employed by the Hopi Indians of the Colorado plateau and Rio

Grande valley in the USA and by Spanish peasant farmers in the Ebro valley in Catalonia in the north-east of Spain. Since the typical features of the terrain on the valley floor consist of a series of discrete channels constantly joining and bifurcating and, in each, a linear arrangement of silts, sands and gravels it is clear that there is a similar pattern of cereal cultivation which may change from one year to the next as the channels and their discrete elements also change.

The discontinuity of the cultivation pattern is also repeated on the alluvial fans which also offer special problems of utilisation when cycles of incision rather than deposition occur. The distinctive process of stream action is here the build-up of the cone form by the seasonal shifting of stream courses scarcely incised in its surface and the problem for the cultivator is to ensure that certain channels will carry the water after the next rainfall. This involves building a system of simple levees to contain the waters within the preferred channels in which the crops are planted when the waters have subsided. From developments such as this it is only one stage further to water diversion by earthen dykes or furrows and with this elaboration of technique one enters that somewhat ill-defined zone between flood-channel farming, and canalised flooding, accomplished as in the hill-foot zone of the Atlas mountains in Tunisia by dams, sluices and diversionary canals which lead water to areas specially selected by virtue of their form and surface materials to facilitate levelling or terracing.

Diversionary canals leading from dams across seasonal streams, or the diversion of seasonal floodwaters into basin areas which the cultivator has selected for their productivity, have been frequently used as by the Hopi Indians and by the ancient civilisations of Egypt. In the Nile valley, winter cultivation of wheat, barley, lentils, onions and flax was possible since the floods were regular and of sufficient amplitude to allow flood-basin farming. Originally this technique simply took advantage of the natural camber of the flood plain created as the river builds up natural levees of coarser material along the main channel but spreads the finer silt at bankful stage into the lower bottom lands. After the floods subsided they could feed stock and cultivate cereals on the still moist fine-textured soils although the seasonal flood hazard restricted permanent settlement to the higher levees and the marginal bluffs. A delta situation offers similar prospects of flood-basin cultivation, with or without diversionary canals, although the cultivated areas have here a more linear arrangement since they coincide with the seasonally filled distributaries. Under these seasonal flood conditions the cultivator avoids the temptation to exhaust the soil by taking several crops in the course of the year and the problems of salt accumulation and waterlogging.

Although it has often been stated that Egypt is the gift of the Nile and its silt, it is also believed that the enforced fallowing imposed by the seasonal flooding was equally important. Fallowing in the low-water period allows the soil to become aerated and the opening up of the soil structure permits a thorough flushing of salts from the soil by the floods which also spread renewed fertility through silt deposition. Sterilisation of land by salination and waterlogging—the pitfalls of perennial irrigations—are thus avoided as the cultivator is forced to conform to the dynamics of his environment. Perennial rather than seasonal irrigation, however, offers so many advantages in terms of yields, diversification of crops for subsistence and industry, extension of the land under cultivation and the expansion of the population that its extension in the dry lands has been restricted only by technical skills, capital resources and the availability of underground or surface-water resources.

With elaboration of techniques, alluvial fans offered the opportunity to convert seasonal run-off into a form of semi-perennial or perennial irrigation which is still widely practised, and allows more permanent cultivation of cereals, legumes and fruit trees. The gradation from coarse to fine materials from the apex of the fan to the periphery encourages seasonal floods to sink into the upper area of the fan, an infiltration which is sometimes encouraged by the erection of simple dams at the lower margin of the coarser material. Through this combination of terrain and technique, ephemeral or seasonal showers are converted into perennial ground-water. In Iran, where the almost complete lack of precipitation between May and October prohibits summer irrigation and where many rivers have a high salt and gypsum load, the inhabitants learnt very early that hill-foot fans offered plentiful supplies of ground-water sufficiently sweet to be tolerated by men, crops and animals. By digging wells in the upper part of the fan—sometimes to depths of about 92 m (300 ft) below the surface—they established the location of the most copious ground-water supplies into which was driven the principal well. From this other wells, separated by distances varying from about 91 to 274 m (300 to 900 ft), were dug to discover the continuity and gradient of the water table and, in turn, these were connected by tunnels called, in Iran, *qanats* or elsewhere *foggaras* or, as in Cyprus, 'chains of wells'. Such wells and underground aqueducts require a very specialised knowledge of ground-water hydrology and are both difficult and expensive to build and maintain. They must also be sufficiently large for maintenance in use so that the water occupies only a small proportion of the cross-section which, at the maximum, is about 1·60 m (5 ft) in diameter. Yet despite the problems of a specialist labour force for construction and maintenance and high capital and recurrent costs, over 40,000 *qanats* still form the basis of irrigation in Iran. On the periphery of the fan, the water is led to the fields and trees, but on the way it is also used to drive water-wheels for the milling of grain. The settlement and cultivation patterns are closely tied to these distinct landforms (fig. 9).

Perennial irrigation from allogenic streams using dams and diversionary canals offer, however, now, as it has in the past, the greatest opportunities for major concentrations of population and the development of the 'hydraulic civilisations'. The simplest technique is to throw a diversion weir across the river and lead the water to the fields; more elaborate techniques involve the building of barrages such as the Delta Barrage immediately downstream from Cairo, the Sukkur Barrage on the Indus, or the Hindiya and Kut Barrages in Mesopotamia. In each case the object is to raise the head of water sufficiently to allow diversion by canals to the lower lying ground. The success of the Delta Barrage on the Nile was such that, as the cultivated area and the population increased, it became necessary, in 1884, to strengthen the barrage to provide more water to cope with increasing demands and, when this proved inadequate, to build dams and reservoirs. The first storage dam on the Nile was built where the river cuts into the granite at the First Cataract. Storage reservoirs are now common on all the major rivers of the dry lands but the Aswan Dam is of especial interest in that it brought half a million acres in lower Egypt under cultivation and, as in the case of the Delta Barrage, supply created further demand which led to a raising of the Aswan Dam in 1912 and again in the mid-1960s, to complete the greatest barrier on the Nile which provides not only irrigation water but hydro-electric power as well. The extent of the use of the Nile waters for irrigation becomes even more striking when it is re-membered that, in 1952, was completed the Edfina Barrage on the Rosetta distributary on the delta, to ensure that the waters of the Mediterranean are prevented from reaching the Nile and the Nile waters prevented from reaching the sea. Such a combination of dams and barrages surely represents a refinement in the provision of water for perennial irrigation for which it is difficult to find an exact parallel in other parts of the world. The major perennial river valleys of the arid lands in the Old World do, however, demonstrate that, as in the case of the Nile, the irrigation works early required a substantial specialised labour force (like the *qanat* constructors of Iran) and that this led to urbanisation, as at Mohenjo Daro and Harappa in the Indus Basin. Here the pattern of technical development was somewhat similar to that of the Nile, proceeding from inundation basin farming to diversionary canals which culminated in 1859 when the Upper Bari Doab canal was dug to link the Ravi with a barrage at Madhopur. By 1917 the system had been elaborated by the construction of the Sidhnai and Lower Chenab canals; later the Grand Triple Project linked the waters of the Jhelum to the Chenab and the Chenab to the Ravi. Perennial irrigation of the lower Indus valley, the Sind, commenced with the building, in 1932, of the Sukkur Barrage followed by the Ghulam Mohammed Barrage near Hyderabad and more recently with the construction of the Guddu Barrage about 160 km (100 miles) upstream from Sukkur.

The irrigation systems discussed so far have all been working under gravity flow whether from seasonal streams, perennial rivers or, as in the case of the *qanats*, from ground-water. Yet some of the earliest principles of mechanical devices were introduced through irrigation techniques for the uplift of water from rivers and wells before gravity flow completed the final transmission of the water to the plots. Such devices range from the ancient Archimedes screw (now constructed of welded steel and competing in cost efficiency with modern pumps), the *shaduf* or bucket suspended from a counterweighted spar, the *noria* or Persian wheels in which water is lifted up in the buckets of slowly revolving wooden wheels to the modern diesel engined or electrically driven pumps. The wells themselves have shown similar developments from hand-dug wells which tap ground-water near the surface to the modern wells of Pakistan and India where perforated steel tubes up to 91 m (300 ft) long are driven into the ground and operated by electrically driven pumps each irrigating up to 324 hectares (800 acres) for cotton, maize, millet and sugar-cane in summer, and wheat and millet in winter. There are nearly 2,000 of these wells in Uttar Pradesh alone. In Queensland water from the Great Artesian Basin is now lifted to the surface by over 2,000 wells with yields of over 250,000 gallons per day.

Uplift of water on such a grand scale contrasts with the smaller quantities handled by horticultural methods in the small oases of Libya. At Jalo the wells are first dug in the sand to a larger diameter than required and then lined with whole or split palm logs to prevent the sides from collapsing. The palm log revetment does not usually extend to the bottom of the well since the sand about 3 m (about 8 to 10 ft) below the surface is cemented by the salts of the percolating waters. From some of these wells water is lifted in leather bags or tin buckets by hand; at others, a depression is excavated to form a ramp down which an animal can walk to lift the water by an ingenious device into a trough. Water from the main wells is led along a series of channels about 15 cm (6 in) deep and the supply to the various branches is controlled by dams of heaped-up sand. For the most part, the various crops are grown within the irrigation channels themselves and the young shoots and seedlings are protected from the sun's rays by small umbrellas of palm fronds (fig. 11). The turnips, onions and tomatoes

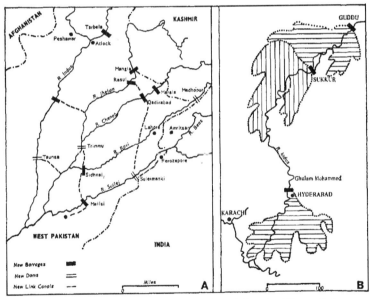

Fig. 10. A – irrigation schemes projected for the Upper Indus Basin. (Based on a map of Messrs Coode & Partners, Civil Engineers)
These are in addition to the long established canal systems of the Land of the Five Rivers.
B – areas irrigated and to be irrigated in Sind. (After Huntings)

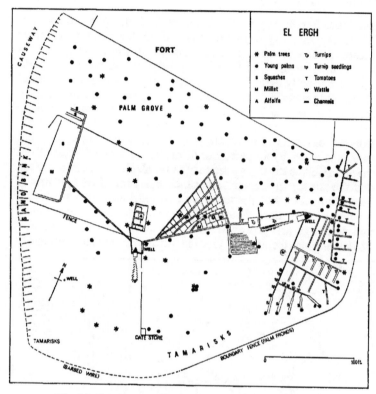

Fig. 11. Specimen garden in the oasis of Jalo, Cyrenaica

receive moisture from every watering which may take place several times a day. Squashes, such as melons and cucumber, millet, barley and alfalfa are grown in small open plots fed by the irrigation channels. These plots are usually divided into small squares by banks of sand where young date palms which require frequent watering are also often situated. Most of these crops are subsistence crops but the tomatoes are for export to the coastal towns of northern Cyrenaica, such as Benghazi, while the alfalfa is for fodder, as is some of the barley. In spite of the constantly high temperatures there is a seasonal rhythm of agricultural activity.

Such cultivation by 'hydroponics', in which water is applied to sand similar to that of the desert proper, lies at the opposite end of the spectrum from the great perennial irrigation cultivation of commercial crops along the Colorado, the Nile or the Indus. The Upper Niger Project clearly shows the difference in scale. Formerly the upper reaches of the Niger discharged into a vast inland sea and the fossil delta forms 809,500 hectares (2,000,000 acres) of potentially productive land if water can be applied. As early as the 1920s the French began a feasibility study but it was not until 1948 that the barrage at Sansanding was completed with diversionary canals from the Niger at a cost of over £10,000,000. By 1949, over 12,141 hectares (30,000 acres) were under cultivation with American medium-staple cotton in the south and Egyptian cotton in the north with, in addition, about 14,170 hectares (35,000 acres) of rice which it is hoped will replace the traditional millet as the principal foodstuff of this and other areas of dry west Africa, and will offer an alternative mode of life to the traditional stock rearing.

Although it would be possible to enlarge on the growing of cotton in the Nile Delta, or in the Sudan where the Gezira scheme based on the Sennar Dam and reservoir allows over one million acres to be irrigated, or in the USSR on the Zeravshan and the Kara Darya in Fergana, one must turn to the south-west of the USA to find the greatest concentration of commercial and industrial crops. From the early settlement by the Mormons of Utah with irrigation techniques which employed Old World devices such as the *noria* the area round the Great Salt Lake has been developed by the building of huge feeder canals into a great orchard with fruits such as apricots and peaches as well as large-scale market gardening providing vegetables for the towns of the eastern seaboard. By 1872 irrigation was being widely experimented with in California, commencing with sugar-beet but later becoming more concentrated on fruits and vegetables. In the Imperial-Coachella valley, whose drainage is to the Salton Sea, development was accelerated by the construction of the All-American Canal in 1942. In the northern (Coachella) section there are now nearly 2,429 hectares (6,000 acres) of dates and over 2,834 hectares (7,000 acres) of a table variety of seedless grapes, while 1,000 hectares (2,500 acres) are given over to citrus fruits, notably grapefruit. Portions of this section are also used for fodder crops such as alfalfa and for irrigated hay and pasture where animals are fattened before despatch to the Los Angeles market. There is a strong concentration on mechanisation of all farming activities and the harvesting is often performed by contractors. Farther south, in the Imperial Valley, where the soils are heavier, there is

E

more concentration on field crops, especially winter harvests of
lettuce, tomatoes, peas, melons and carrots which are transported
in insulated lorries and railway vans to the east. In addition to
these 25,900 hectares (64,000 acres) of truck crops, there are also
11,331 hectares (28,000 acres) of cotton and 13,759 hectares
(34,000 acres) of sugar-beet refined locally or in the coastal area.

Perennial irrigation on these scales involves massive capital
investment in reservoirs and canals and is impossible without
government aid. With it was built the complex system of dams on
the Colorado River and its tributaries, the product of the period
of reclamation which reached its climax in the 1930s but which
had begun at the turn of the century somewhat belatedly after the
recommendations of J. W. Powell. The first dam near Yuma was
followed by others such as the Hoover, Parker and Davis Dams
on the Colorado in the Mohave desert, the Gillespie and the
Coolidge Dams on the Gila, the Horseshoe and Bartlett Dams on
the Verde and the Roosevelt Dam on the Salt River. The Phoenix
'oasis' in Arizona receives water from the Roosevelt Dam 188 km
(117 miles) away which irrigates over 7,689 hectares (19,000 acres)
for dates and olives, oranges and grapes, as well as cotton and
lucerne. In Australia the pattern of irrigation development has
followed somewhat similar lines with, for instance, the setting up
of the River Murray Commission in 1915 from which has resulted
the creation of six major storage reservoirs for flood control and
irrigation. More recently the Snowy River scheme has been
created to transfer water to the Murray and to generate hydro-
electricity. Examples from India such as the Thar Development
Authority set up in 1949 similarly demonstrate that national
government schemes are necessary, as they have been in Egypt,
for such large-scale development, and that for the developing
countries international financial aid is essential.

Although perennial irrigation has undoubtedly had a beneficial
effect on the economies of the countries of the dry lands and has
permitted increased population and standards of living to levels
well above those which could be achieved by dry farming or
seasonal irrigation, it has also created environmental problems
and in some areas reduced the area of land formerly cultivated. If
water can make the desert bloom it can also sterilise the land as
effectively as does wind erosion under poor techniques of dry
farming. With apparently unlimited supplies of water too much
is applied to the fields and under conditions of poor drainage the
ground-water table rises and formerly cultivated land is replaced

by lake and marsh, as in parts of the Nile Delta and in the Sind. Drainage canals and pumping stations must be provided at great expense to maintain the correct ground water-level and are essential to restore formerly irrigated areas to productive use. It has been proved that both irrigation canals and drainage canals must be built and integrated with the levels of the fields if perennial irrigation is to continue to operate without harmful waterlogging of the soil, and the systems of cropping must also be carefully controlled. A return to the old water-lift devices has been encouraged to reduce the amount of water applied to the fields.

Waterlogging is often accompanied by an increase of salts in the soils to levels which even the most salt-tolerant crops cannot withstand. Much water from perennial streams and from ground-water contains large quantities of salts so that, as it is spread over the fields, the salt content of the soils increase. Some arid-land soils are saline such as the *solonchaks* and *solonetz* (see pp. 79–80), while the rise of saline ground-water can also sterilise large areas. Salt efflorescence may spread over the surface and the salt particles be blown by the wind to areas which otherwise would be salt-free. Flushing of the soil and drainage are remedial measures which have been undertaken in the Sind and the Nile croplands; in the latter, rotation of irrigation and drainage has been employed in the same canal system. To avoid both waterlogging and salinity the correct utilisation of the micro-relief of these areas of gentle slope becomes imperative.

The building of storage reservoirs is accompanied by a reduction in the amount of silt spread over the fields and by the build-up of deposits within the reservoir itself; the first leads to a loss of fertility, often aggravated by over-cropping, since the silt is no longer spread over the fields as in flood farming. The soils require artificial fertilisers, the cost of importing which must be set against, as in Egypt, the value of the exported cotton or the subsistence crops. To maintain the volume of water stored behind the high dams used for perennial irrigation expensive dredging operations are necessary while the canals themselves require cleaning.

The changes which result from manipulation of the natural landscapes in the perennially irrigated lands also have important side-effects on the insect and parasite population which thrive under the favourable moisture conditions provided by irrigation and have important effects on the health and socio-economic life of the populations. Insect- and parasite-borne diseases are most pronounced in Old World irrigated lands, since in the American and Australian deserts the standards of public health are higher and both remedial and preventive medicines are more readily available. Where the irrigated areas coincide with dense populations at subsistence level it is difficult to control the endemic diseases which are related to flowing and free-standing water in a hot climate. Malaria is endemic in most of the Old World oases and has been a major problem since irrigation farming was first attempted in the riverine lands and deltas; bilharziasis, which causes intestinal and liver diseases, is of equally long standing, since calcified eggs of the species of worms which affects the urinary system have been discovered in the kidneys of Egyptian mummies. Both the Anopheles mosquito, which transmits human malaria, and the schistosomes, or blood flukes, which are parasitic in man, produce debilitating diseases which have been shown to ruin the health of a country and to render complete irrigation schemes, such as one established in Southern Rhodesia after the Second World War, complete failures. Problems of irrigation and public health thus go hand in hand and the development of one without the other can only lead to economic and social disaster.

There are assuredly many lessons to be learnt from the problems of perennial irrigation of which probably the most important is that with increasingly high standards of technology there has been a tendency to ignore the fundamental principles of conforming with the environmental opportunities rather than allowing the techniques of control and supply of water to be applied without discrimination. When crops were grown on a seasonal basis as water and climate permitted and when villages were sited on the marginal bluffs at water-supply points, problems of waterlogging, salinisation, overcropping and health problems were less pronounced and almost self-remedial. It is now being realised that the almost universal perennial irrigation systems expressed in huge dams and major canals are wrong in principle, since they lead to silting, high rates of evaporation and loss of water. The water is wasted, leads to major land-use problems and in turn affects public health. One is almost forced to conclude that the *qanats* of Iran offer the best of both worlds, with underground storage fed by seasonal and perennial run-off. The most fruitful forms of perennial irrigation must, in the future, be more carefully devised to fit in with, rather than be built on to, the physical environment.

7

PASTORALISM—A BASIC DRY-LAND RESPONSE

The green spots and strips of irrigation cultivation seen on a satellite photograph of the arid lands occupy a much smaller area than other types of land-use. Dry farming has been shown to fill in some of the gaps but by far the most ubiquitous economic response to dry lands is pastoralism. Pastoralism, however, embraces many ways of life and standards of living. At one end of the scale is the wealthy rancher of cattle or sheep in the USA or Australia who produces beef and hides, mutton and wool as a commercial enterprise with a title to hundreds of square miles worked from a permanent base and providing paid employment for a few tens of people. This way of life is far removed from that of the nomads of the Old World whose objective is subsistence for themselves and their families. There are, in fact, as many different forms of pastoralism as there are methods of cultivation in the arid lands. Pastoralism is also a way of life which has been subject to rapid changes in scale and distribution, changes which have been initiated by changes in the environment, as through climatic change or rainfall variability, or through social, economic and government pressures.

Evidence from the Old World points to south-west Asia as the area in which animals were first domesticated although as yet there is no absolute agreement as to the ways in which this was achieved. Some believe that in areas where game was relatively scarce, as in most of the arid lands, there was an incentive to capture and to keep the animals alive until they were required for food. Others prefer the concept that, in the period of desiccation which followed the pluvials of the glacial periods, man and animals were drawn together by common need at water points and that from this unity of interest in survival there was the opportunity for domestication. In more humid areas where game was plentiful quite large communities could live above day-to-day subsistence level and have the leisure to elaborate the arts depicted by the prehistoric inhabitants of the caves at Lascaux in France, at Altamira in Spain or by the rock paintings of the Ahaggar in the Sahara. In such localities and under such climates there was less incentive to domesticate since a ready supply of comparatively slow-moving animals was available provided that population increase did not outstrip local food supplies.

Discoveries of archaeologists give clues as to the diversity of approach to the problem by primitive man. They show that the hunting of animals was not indiscriminate. Neanderthal man living in caves over 40,000 years ago at Shanidar in Iraq seems to have acquired some specialisation in the species he hunted. Onagers were available on the plains at the hill-foot and goats in the hills, yet the 'kitchen midden' bones are those of the goat rather than the onager. Recent discoveries at Ali Koosh in upper Khuzistan in south-west Iran show that at an early stage of occupation diverse animals were hunted with no evidence of domestication. Later, however, goats were domesticated as is shown by changes in the shape of the horn core which is paralleled by other physical changes in the bone structures of animals after domestication. On the evidence available it would appear that there was local or, perhaps, regional specialisation which through cultural diffusion, trade and the exchange of animals (and the knowledge of their breeding) led to a spread of the skills of domestication over a wide area of south-west Asia, a development accompanied by the establishment of permanent or semi-permanent settlements in areas where fodder was plentiful and assured, either through favourable seasonal rainfall or a high ground-water level. Such a settlement has been excavated at Jarmo in north-eastern Iraq and shows that by 6750 B.C. goats, pigs, dogs and possibly sheep had been domesticated. From Shanidar, farther to the north, a Carbon-14 date of 8800 B.C. for domesticated sheep is available. These are late dates compared with those for the scavengers, the dogs who from Palaeolithic times linked themselves in a symbiosis with the wandering bands of hunters.

The early domesticators of animals were faced with the same problems as the cultivators—the necessity to use the environmental resources the seasonal rhythm of which is controlled by the periods of rainfall. It is reasonable to assume that there was a degree of coincidence of the landforms and water supplies used by both ways of life since, where plentiful forage was available, propitious circumstances for cultivation were likely. In a seasonal rainfall climate, i.e. in the semi-arid rather than the extremely arid areas, the pasturage varies in an annual but nevertheless regular cycle and the pastoralist can be fairly certain that the same area will be available from one year to the next. Within the area there will be, however, a diversification of terrain ranging between the floodplains, channels and alluvial fans which can only be used when the floods have subsided, and the interfluves, on which the quality of the pasture will depend on the amount of rainfall. These pastures will be available before those of the watercourses but will not last as long since the grasses will soon exhaust the available soil moisture. Superimposed on these problems of pasture, however, is the need for regular watering of the stock so that the pastoralist cannot stray too far away from the regular waterholes with sheep, cattle or goats although the camel offers a wider range of forage exploitation. He must also ensure that in his cycle of areal activity there is the opportunity for the animals to include mineral elements in their diet derived either from pastures growing on a wide variety of soils or by periodic visits to salt licks. This limited cycle of migration is roughly equivalent to the move from low to high pastures in the Alps, the Carpathians and in Scandinavia, and may, likewise, be termed *transhumance*. It is a rational adaptation to the environmental resources and corresponds to the flood-channel or flood-basin farming of the cultivators. It avoids the concentration of animals and people in high-temperature/high-humidity conditions the whole year round and thus reduces the risk of insect-borne and parasitic diseases both for men and for stock. A major problem, however, is the concentration of too many animals in areas of abundant waterholes since although there may be ample water to drink there is insufficient pasture to maintain large herds. The distribution of this transhumant pastoralism is related to certain climatic zones such as the Sahel and the Sudan to the south of the Sahara but there is no direct correspondence since the tribes may overlap their area of activities into other bio-climatic zones as their experience and cultural ties may determine. The transhumant principle is, of course, not confined to pastoralism in the developing countries since it is also employed on the very large ranches of the United

States and Australia where, however, under the best conditions of management the stocking rates are more carefully controlled.

Transhumant pastoralism, often associated with cultivation of cereals and fodder crops, as in the Tell of North Africa, may thus be explained as the rational response to environment at all levels of culture and technology. It is far more difficult to explain the reasons for nomadic pastoralism, the way of life which is often regarded as the characteristic of the deserts. Why should pastoralists forsake the security of the semi-arid seasonal rainfall areas and follow a more precarious existence of almost constant mobility in areas of sporadic rainfall whose distribution and amount will vary so markedly from one year to another? Men have demonstrated the world over that individual needs for shelter, defence, food supply and reproduction are best satisfied by congregating in those zones where natural resources permit higher population densities. When so many people throughout history have shown a desire to collect material possessions to increase their personal comfort, the skin tent and a minimum of household utensils and furniture seems to be at variance with basic human desires. Why should the nomadic way of life have developed and why has it persisted for so long, especially in the arid regions of the Old World, when the end is to serve the animals from which not all men's needs may be obtained? The nomad must have dates, grain, sugar, tea, coffee and cloth, as well as milk, meat, hides and hair.

There have been many answers to these questions but no one answer seems sufficient in itself. In 1935 Toynbee regarded the increasing challenge of desiccation as the principal factor in the development of nomadic pastoralism whether from people who were originally cultivators or transhumants. Some were assumed to have moved to areas of greater humidity (the evidence for this is far from adequate) but others were believed to have accepted the more severe challenge of the periodic rather than the seasonal rainfall areas and moved with their herds into zones where migration routes were defined by water-holes in association with unpredictable pastures. They were inspired, perhaps, by the belief that ways of life which involved the domestication of animals were superior to cultivation—a domestication in which one might differentiate between that which involves training, such as in the case of the camel, horse or dog, or without training such as for sheep, goats and cattle.

Other reasons can, however, be put forward for the acceptance of pastoral nomadism as a way of life. Population increase may have led to overpopulation so that families and kinship groups were forced to move away, while the pressure of population in a small area of transhumance would have induced environmental changes through the anthropic element in landscape change such as overstocking, depletion of pastures and soil erosion together with the changes in vegetation which result. The difficulties which arise when trying to explain nomadism are, however, partly resolved when it is remembered that this way of life is not isolated from the other forms of land-use and societies in the arid lands but, in fact, is complementary to them—that there is a symbiosis between the nomadic pastoralist on the one hand and the cultivator who exploits either dry farming or oasis irrigation techniques.

This relationship is in part induced by the need for exchange of products between the two economies and in part the product of the greater mobility and hence fighting power of the nomad who, in the individual tribe at least, has a leader who has been selected for his ability to take decisions on problems of water supply or pastures or for his ability to lead in battles for grazing rights or for plunder. The cultivators needed milk, meat, hides and wool as well as grain. In areas of winter rainfall such as eastern Syria, Jordan and the northern parts of Arabia, the farmers began to cultivate in the autumn at a time when the grasses are beginning to spring up in discontinuous areas of the desert between the cultivated lands. The Bedouin moved away from the settlements until after the harvest when the nomads began to re-enter the cultivated areas to graze the animals on the stubble so that from July to October nomad and sedentary cultivator were in close proximity. Regular links of this kind may be extended into a more intimate economic connection between the desert and the sown in which the same families return, according to the seasonal rainfall pattern, to the lands or palm groves which they have themselves cultivated or own. The Doui-Menia of the Sahel both grow grain and own date plantations while the Teda, the Negro nomads of the Tibesti, own plantations in the oasis without cultivating them as do some Bedouin in the oases of Jalo and Aujila in the Libyan desert. At the time of the date harvest their black tents may be seen clustered on the margins of the oasis. The Tedas and others used slaves to cultivate the date palms and since the abolition of slavery the work is done by former slaves, the Kamayas, who are still linked economically and socially with their former masters. Such a blend of economic activity serves to distinguish two principal types of nomadism, the one found on the margins of the extreme deserts in areas of seasonal rainfall, or in the pluvial upland areas, typified by the Teda, some Bedouin tribes, the Gherib of Tunisia who may be classified as *semi-nomads*; the others are the *true nomads* whose contact with the oases and steppe margin is irregular and who depend for their existence on exchange and trade. There is, of course, no clear-cut classification especially today when the nomadic way of life is in a state of transition.

The true nomads are relatively few, less than 750,000 in the Arab lands of the Middle East, whereas there are probably more than 2,000,000 semi-nomads in the same area. The size of their social groupings is self-regulating since as the pressure on grazing increases they are forced for survival to split into smaller groups which tends to perpetuate tribal and sub-tribal conflict since they all demand the same natural resources—grass and water. The migrations of the nomadic groups are controlled by the availability of these resources which unfortunately do not always coincide since exhaustion of water at the drinking holes may occur while there is still abundant pasture in the vicinity—'vicinity', of course, will vary with the types of animals which are herded but for the camel the limit is as much as 48 km (30 miles). Even though the pastures may appear in different places from one year to the next, their wanderings follow a fairly fixed annual cycle determined by water-holes and landmarks within boundaries which are the result of historical demarcation disputes resolved by fighting. Such boundaries are by no means permanent since any sign of weakness will be exploited by the nomads of the neighbouring area, so fierce is the competition for pasture and water.

Such wanderings have paid scant attention to international boundaries and have given rise to friction between states until treaties were established. From a wide range of international agreements those in the Middle East and Africa may be cited as examples. The Ankara Accord between France and Turkey in 1921 permitted the nomads to move between Syria and Turkey without the payment of taxes. By the Treaty of Ankara in 1926 provisions were made for the maintenance of neighbourly relations on the frontier north of Mosul between the United Kingdom, Turkey and Iraq. Italy and Egypt came to an agreement in 1925 for the regulation of nomadic affairs and to provide for the use of wells on or near the Libyan frontier while, in 1924, Great

Britain and France agreed that the Wadi Howa, the boundary between the Anglo-Egyptian Sudan and French West Africa, should be used in common by the tribes living on either bank. International agreements such as these simply recognise the geographical basis and the dynamics of societies whose economy is controlled by environmental factors which international boundaries ignore. They are comparable with the disputes which have arisen through the use for irrigation of perennial rivers which flow through several countries.

The migratory habits impose numerous restrictions and would have been impossible without the goat-hide water-bags or, more recently, the jerrican. The size of the tent and the amount of furniture are limited by their portability, as the modern car camper is well aware. When capital is in stock rather than in money it is difficult to enforce discipline and, as T. E. Lawrence discovered, even the provision of money did not automatically solve the problems of tribal cooperation. Laws tend to be concerned with the principle of an 'eye for an eye and a tooth for a tooth' since there can be no prisons and not much power in public opinion. Some would also regard herding as an occupation which gives the maximum opportunity for self-contemplation and speculative philosophy assisted no doubt by the impressiveness of the firmament viewed from the cool sand in the desert night.

The situation of the Old World Deserts between the potential trading areas of the tropical savanna and humid rain forests in the south, or the riches of the Far East and the urban civilisations of the Mediterranean and the lands to the north, provided an alternative form of nomadism derived from, and often linked with, pastoral nomadism and subject to similar environmental controls. There are no real parallels for the nomads' trading functions in southern Africa, Australia or the Americas where exchanges were made within rather than with the peripheries, but *commercial nomadism* has long been in the Old World an important and lucrative form of employment, to which the opportunities for pillage were an added inducement. From simply transporting water, tents and the products of the oases for his own use, the nomad extended his sphere of activity and began to transport for others, developing a haulage system based not only on the camel, which was used principally in the very arid areas, but also on oxen, horses, asses and mules as well. It was common, for instance, in the trans-Saharan trade for goods to be transferred from camels to oxen on the semi-arid margins and this specialisation in the types of animals used for transport was also paralleled by concentration on certain types of goods. The Kel-Oui—a branch of the Touareg, in the south of the Sahara—for instance, take millet and cotton cloths from the Sudan to Bilma near Lake Chad and return with salt. Some Saharan caravans concentrated on slaves obtained in profusion at desert ports such as Timbuktu and Gao. Without the commercial nomad and his caravans there could have been no spice and silk link between the Far East and medieval Europe. The deserts which separate the Mediterranean from the Indian Ocean would have remained for a longer period the barrier they appear to be on the bio-geographic map. Nor must the role of the commercial nomad in bringing the pilgrims to Mecca be forgotten—was not Mohammed himself a camel boy with a caravan? Commercial nomadism is, however, becoming more and more a thing of the past. The value of the camel has declined—in Tagant (Mauritania) a camel in 1910 was worth twenty milch cows, today it is only worth one; the 15,000 camels which formed the caravan between Timbuktu and Taoudeni are no longer required. New methods of transport, desert oil, the opening up of the sea routes, the suppression of the slave trade and the ever present problem of safety have all contributed to the reduction of the caravan trade and its predators. The Touareg has now become a policeman.

The decline of commercial nomadism is also, of course, related to the decline in pastoral nomadism. Almost everywhere the tendency is for the sedentarisation of the nomad whose poverty has been accentuated by the extension of cultivation to areas which previously were grazed by the nomads' flocks and by the protection afforded by the civil power to the former 'protectors' of the oases and caravans. The movement towards a sedentary life was probably generated within the nomads themselves as over-population built up internal pressures and forced the weaker groups away from the central core of the nomadic world. Driven, as in Arabia, into peaceful contact with the cultivators, they became sheep-herders who found that contact with the towns and villages of Syria and Iraq, themselves reflecting the rule of alien civilisations such as the Romans, Europeans or Turks, and life near permanent water with a ready market for sheep, wool, mutton and milk in the towns was more attractive to those who had previously known only the rigours and violence of the nomadic existence. They could, however, still live and work with the animals which all pastoralists the world over come to love.

Elsewhere pacification, as by the French in the Sahara, converted the true nomad into the semi-nomad and both have been transformed into sedentary ways of life. It is true that there have been reversals of this process resulting from particular social and political circumstances, but the countries of the arid lands are faced increasingly with the problem of assimilation, education and the methods of encouraging the permanent settlement of nomadic and semi-nomadic tribes. Improvements in the infrastructure of the arid lands of the Old World have offered many opportunities for employment on roads, railways, canals and irrigation projects but there is little doubt that it has been the discovery of oil which has been by far the most potent force, not only because it has offered direct employment but because it has brought contacts with the outside world. Such developments have had mainly an indirect effect but Egypt, Syria and Iraq in particular have initiated programmes specially designed to settle the nomads, a course which has been facilitated by major land reforms which have reduced the areas which individual landowners can hold and have made available irrigated lands both for the nomads and for the share-croppers of the old estates. Since the nomads can find a place only at the lower levels of sedentary society, programmes for technical education are urgently required otherwise they will simply swell the ranks of the impoverished groups of these dry-land societies. Each country has different social, political and environmental problems and for each the solution to the problem must be different but it would be wrong if the inherited stockbreeding skills of these nomadic peoples were to be ignored in areas where the environment beckons pastoralism.

Stock-rearing by primitive peoples in the past and at the present day has always been based on conservation of the animals rather than slaughtering them for meat. Killing the animals destroys the capital asset, and thereby the wealth of the individual or family, in return for short-term meat products; for many nomadic peoples slaughter could be countenanced only when the animals became infirm, injured or too old to keep up with the migrating herds. Pastoral peoples such as the Hausa and Fulani of West Africa, or the Somalis, Turkana and Masai of East African arid areas, currently depend on diets of blood and milk. The animals act as filters for water which is too saline to be drunk by man or in which the excess fluorides would have a deleterious effect on health. Moreover, as Pearsall has pointed out, 'the use of milk and blood greatly reduces the large conversion losses of these (dry) habitats. Whereas the flesh of an animal represents at best about one-tenth of the protein eaten, milk protein is equivalent to about one-quarter of the protein used in producing it, and the blood probably represents a similar economy.'[1] At the same time the waste of much of the animal carcase (although primitive peoples, or peoples accustomed to food shortages, consume far more of the animal than does his modern counterpart in an urban civilisation) means that not much more than 40 per cent of the weight of the beast is eaten. The Old World sheep herder was, and is, more interested in the wool than the lamb and mutton and, although sheep, goats, camels and cattle were bartered for the products of the oases, the total number of animals involved was very small in comparison with the total in the herds. That the size of the herds is often in excess of what the pastures can support has been shown in areas where the pressure of grazing by domesticated animals has caused progressive deterioration of the habitat. Under natural conditions years of drought would have meant a reduction in the size of the herds, to a figure estimated to be roughly the equivalent of one-tenth of the dry weight of the vegetation.

The penetration of Europeans into the dry grasslands of the Americas, Africa south of the Equator, and Australia has produced a rather different pastoral response to that of the Old World. In many new lands pastoralism preceded cultivation as the new immigrants brought their domesticated animals on to the relatively virgin grasslands of western North America, South Africa and Australia. Here the grazing of animals in areas of very low population density was easy since they did not have to compete for land with the already established farmer. The reversal of this process has led to much conflict in land-use between the grazier and the cultivator of cereals which, in North America at least, has been written into the folk-history of the United States. A second major difference is that many animals were grazed and reared not for their continuing products of milk, hair and wool but for the meat which was exported either on the hoof or in the carcase to the developing urban centres of the new lands and later, with canning and refrigeration, to the great meat importing countries of the Old World. Another basic difference is that although grazing managed by Europeans is, to a certain extent, migratory, around a centrally based ranch established at a water-supply point, it is certainly not nomadic although it could

W. H. Pearsall, 'Survival in Drought', *New Scientist* (Nov., 1961), **23**, p. 491.

often reasonably be called transhumant. There was no nomadic pastoralism in the New World before the arrival of the Spaniards who brought with them the horse, essential in the herding of sheep and cattle over ranges of such vast extent. Few grazing grounds are as productive as the winter-rainfall areas of the desert grasslands of south-east Arizona yet they can carry an average of only fifteen head of cattle per square mile (compare the stocking density of 300 head of sheep per square mile in the uplands of Scotland under the control of one shepherd).

This low-stocking density is the real link between the Old World nomadic pastoralist and his European counterpart in the dry grasslands. It has been shown that seasonal migration and transhumance are a basic response to the seasonality of plant growth governed in the hot semi-arid lands by rainfall rather than by temperature, while conditions are even more complex in areas with episodic rather than seasonal rainfall. The stock-carrying capacity of the dry grasslands is thus related to the lowest level of plant growth which occurs at the driest time of the year. In the Tropics it has been said that the same number of hectares are required to graze one ox as there are months in the dry season but this assumes that there is no migration or transhumance and no scientific management of the pastures. Much can be done to conserve grass, shrubs and water but conservation implies a high standard of range management which the European pastoralist took a long time to learn in the USA, Africa and Australia. Formerly the only answer to deterioration of the pastures was migration especially before the fencing of the ranges but now it is possible to rotate pastures to conserve fodder and moisture while more modern concepts envisage the growing of fodder in cultivated areas, often under irrigation, in close proximity to the pastoral zones.

Cattle and sheep and, at an earlier period, horses, have been the main stock reared by Europeans in the New World, Africa and Australia. Whether these are the best animals to convert grass into human food in the dry lands is debatable but their selection in the semi-arid lands is understandable since they were the animals with which the European colonisers were best acquainted; given other colonisers, the animals might have been yaks or llamas. The main problem in all successful pastoral activities is to select the animal with the best adaptation to the available pastures and the introduction of European breeds has been most successful where the environmental conditions have been most similar to those of the more humid lands from which they were derived. In many dry lands, however, far better results, with higher yields of meat, milk or wool, have been achieved either by selective breeding of native animals which often have a much higher resistance to endemic diseases or by the crossing of such breeds with those from Europe which have better conversion capacities for the transformation of grass into food—i.e. those with higher productivity rates. Typical examples have been the crossing of European with humped Zebu cattle in East Africa or the mating of European sheep such as the Spanish merino with the native fat-tailed sheep of South Africa. In North America, the rangy Texas longhorn, able to survive on a minimum of moisture and grass, produced but small quantities of meat of rather low quality but it was able to withstand the rigours of trail droving sometimes for distances of over 1,600 km (1,000 miles); nevertheless, by the time it reached the railhead, the beast was in very poor condition. Crossing with Herefords improved the quality of the animals but could not have been commercially successful without the extension of the railway network of the USA which eliminated the very long drives. More recently there have been crossings of the Herefords with the American bison in an attempt to produce an animal which can withstand the biting cold waves which sweep down the bare expanses of grazing lands on the more humid margins east of the Rockies.

Often the problem has been the choice of animal rather than breed and conflicts between the sheep-herders and the cattle-ranchers were often similar to the conflict between the rancher and the dry-land farmer. Sheep are probably more adaptable than cattle to dry climates since they require less food for each animal and are better able to make use of the sparse vegetation. On the other hand it was claimed, especially by the open-range ranchers of the USA, that they were more destructive to the grazings since they pulled up the grass and the roots rather than eating the grass alone and thereby exposed the ground to aeolian erosion. There was undoubtedly prejudice against the sheep, perhaps in part because mutton was less desirable than steak to the meat-eating American, and partly because perhaps the sheep was more slow-moving in the droving era. Pastoral prejudice also, of course, extended to the crop farmer who followed the grazier into the dry lands of western America and began to fence in his land to keep out the migrating herds of cattle from the fields and from the water point on which the homesteader was based. With the elimination of the water point, vast areas of grazing land were rendered unusable by the stock man who often had no clear title to the land across which his cattle had roamed. This conflict of farmer and rancher is well known and understandable in the context of the development of the American west at that period, but it is, of course, the complete antithesis of the relative co-operation of the Old World nomad and the oasis cultivator. Now, of course, irrigation agriculture for fodder crops such as lucerne provides the symbiosis on which all successful dry-land agriculture must depend.

In North America it was the cattleman who pioneered the advance of the grazier into the arid south-west and filled in the wide open spaces crossed by the migrants seeking their fortunes from the gold first discovered in the tail-race of the mill at Sutters Fort in California in 1848. The overland trails led through lands which were the home of migrating Indian tribes who hunted the bison which grazed on the empty grasslands. By the 1860s the pastoralists were beginning to move in from the more humid areas and extended their activities and numbers as the railways followed them to the west. Yet in southern Africa and in Australia it was the sheepmen rather than the cattle-rancher who pioneered the extension of this economic activity into the arid areas.

Stimulus to the development of pastoralism in the drier areas of southern Africa arose from the establishment of the victualling station at Table Bay by the Dutch in 1652 to service the sailing ships for which the Cape of Good Hope was a halfway point on journeys to and from the East. From the earlier grain and wine farming in the winter—rainfall districts within wagon-transporting distance of the harbour, the colonists turned increasingly to stock-breeding and to the pastures which lay to the north in areas where the Hottentots offered but weak resistance to the advance of the pioneers. With the increased demand for meat for troop-ships and warships, especially after about 1730, the colonists expanded their stock-raising activities on the Karoo where the grazing for sheep was so good that the fat-tailed sheep could be driven over 241 km (150 miles) to Cape Town and still arrive in prime fat condition for export to foreign markets in foreign bottoms, assisted by the Hottentots who, unlike the Indians of western North America, rapidly became integrated into the economic activities of the European colonists. There was a much more migratory existence for the grazier than in North America and many sheep-herders spent their lives with their families in

F

covered wagons leading an almost nomadic existence following the grazings as they appeared with the rains. In contrast to the North American grazings, those of southern Africa had to contend with other major hazards—the migrations of huge numbers of wild animals (estimates place the numbers of spring-bok in hundreds of thousands) which, driven by drought from their usual pastures, intruded on to the lower Karoo destroying all before them as did the swarms of locusts with which they had also to contend.

From these early beginnings in a difficult dry-land environment the grazing industry was changed by the beginning of the eighteenth century with the introduction of the Escurial Merino from Spain and with the occupation by the British in 1795. Pure and cross-bred merino transformed what had been a mutton and tallow industry into a wool-producing industry as well, since the merino was a dual-purpose animal which provided the basis of a major export trade in a form which could stand the long journey to the markets of Europe. Here also the goat became a major object of the grazing economy, partly because it could utilise the drier areas, and partly because the flesh was valuable in an area where game was becoming increasingly scarce. Moreover the goat could be used to lead the flocks of sheep on migrations and it also offered a valuable export product in its skin (Cape kid). With the introduction of the Angora goat and the development of the mohair industry in Britain the future of this animal was assured although the ostrich, whose feathers were to become less in demand with changes in fashion, has not survived as a major herding activity. Nevertheless the demands made by the ostrich for fodder (lucerne) led to the development of many small-scale irrigation projects which have since nurtured important cattle pastures. The record of pastoral activities in the dry areas of southern Africa is thus impressive not only in its scale but especi-ally in the variety of creatures which were reared. Sheep, goats, cattle, oxen, horses and ostriches demonstrate a much wider range than the cattle and sheep of North America and indicate what might be achieved in the arid lands of the northern part of Africa and the Middle East.

For the early immigrant to Australia conditions were very different from those of the American spreading to the west or of the Cape colonist moving his pastoral activities towards the north. Of all the dry-land environments that of Australia was the one least modified by human beings while the native game was far less both in species and numbers. When Governor Philip's convict fleet sailed into Sydney Harbour in 1788 the primary objective was subsistence in the better watered areas immediately around the inlet and it was not until about 1820 that the first major movements into the drier areas beyond the Great Divide took place at a time when the demand for wool in Britain was high and economic depression on the land in Britain, following the Napoleonic wars, encouraged families of free men to follow the earlier convict settlers. With none of the markets of the Cape Colonists for mutton the concentration was obviously on the production of wool; cattle were less important since apart from their use as draught animals there were no large markets for meat products, apart from tallow and hides. Merinos shipped from the Cape in 1796 were to become the mainstay of the grazing industry in Australia but were limited at first by their demands for grazing within a few miles of water which became increasingly scarce to the west although the grazings were good. It was not until the exploitation of the artesian basins began about 1880 that the drier areas could be successfully exploited for sheep-farming, by which time the rabbits, which had been originally introduced to augment food supplies, had got so out of hand that they competed with drought as a prime factor in reducing the amount of grazing land available. Migratory systems of grazing, within the limits of the water supplies, have also been adopted in Australia but have been more used for cattle than for sheep which cannot so readily be transferred from one district to another to follow the herbage of the short-lived annual plants which follow the rain-showers and the more permanently available spinifex and salt bush.

The grazing of cattle in Australia originated with animals imported from Bengal to feed the soldiers and convicts at Port Philip but spread into the drier interior as the frontier of settlement advanced beyond the more humid eastern areas. Beef cattle offered certain advantages once the demand for meat in Australia increased, as with the gold-mining camps, and when refrigeration offered a solution to the problem of distance from the world's markets, but progress in the cattle industry has suffered from competition with beef from the Argentine and it is the sheep station covering over 20,235 hectares (50,000 acres) and the value of the wool clip which dominates pastoral activity in the dry lands of Australia.

Throughout the arid and semi-arid lands the pastoralist, whether Old World nomad or semi-nomad, North American rancher or South African or Australian sheep-herder, is faced with the problem of adjustment to the varying carrying capacities of his land. High variability of rainfall and herbage offer a constant problem of adjustment either by reducing the amount of stock in the poor years or by migration to areas of better grazing. Two solutions seem to be possible and both have been used where economic and technical resources are available; either the animals can be moved by modern transport to the fodder or the fodder can be brought to the animals. Solutions such as these imply the application of scientific resources to the problems of the arid lands, the building up of an infrastructure of roads and railways and to developments in the economy which are more normally associated with the humid areas. Increasingly, forms of economic exploitation of the dry lands other than farming and pastoralism are being developed and the horizons which they reveal are fundamentally different from those so far considered.

8

THE FUTURE OF THE DRY LANDS

The resources of science and technology have been used to solve environmental problems in the dry lands since the beginnings of prehistory and there is no reason to doubt that there will be an intensification of the process as living space becomes more restricted in the humid zone. Already irrigation has been extended over wide areas which had previously been untouched by the cultivator's plough, and reintroduced to zones where invasions and the breakdown of government and society within the historic period had led to a reversion to the dry-land norm. Improvements in the management of pastoral resources in the semi-arid lands permit increased numbers of grazing animals which can be watered from deep wells and bores. The railway and the motor lorry have speeded the transport of goods and people over dry desert tracts crossed previously only by slow-moving caravans and wagon trains; sand yachts have crossed the Sahara and Route 66 gives access across the dry south-west to California. The application of technology to traditional methods of land-use and transport, to housing and urbanisation, has increased the economic value of the 'Dry Third' and there seem to be no insuperable difficulties to making the desert bloom especially where fossil soils are concealed by a thin veneer of sand or gravel. If water is available from perennial rivers, or stored in underground reservoirs, or behind high dams, or desalination costs are reduced still further, then agricultural, or even forestry, productivity seems assured. Large-scale commercial agriculture and pastoralism would appear to be almost universally viable, at least in the semi-arid lands, and have frequently been suggested as possible solutions to world food problems.

Yet increasingly one sees other methods of using the land and water resources which indicate that there might be alternative forms of exploitation to modernised but traditional methods. The extraction of oil and natural gas from beneath the desert surface in North Africa and the Middle East has transformed the economy of those countries fortunate enough to possess such resources since foreign capital is readily available even to the poorest community and technical aid freely given. The process of change in such countries is reminiscent of the ages of technical development in the arid south-west of the USA, in arid western South America and Australia. From the Sahara and Libyan deserts, from dry Iran and Iraq, pipelines transport oil and natural gas to coastal tanker terminals where skyscrapers dominate the dusty coasts. The needs of military technology force the location of proving grounds for aircraft, rockets and missiles to the sparsely populated areas and to the dry open plains with thin or absent cloud cover rather than to the cloudy humid forest lands. The sun-seeking tourist uses the desert for recreation and retirement aided by the provision of air-conditioned motels, houses, cities and filling stations with good metalled roads which feed to airports linking the populated humid zones to the still scattered communities of the arid areas of the continents. Yet it is the development of these isolated centres of population to urban status which is the continuing story of the occupancy of the arid lands and which links the Old World with the New. In the Old World (excluding, of course, Australia) urban civilisations have been a feature of the landscape since the use of the environment reached the stage when food surplus and differentiation of labour allowed large communities to live at favourable water supply points—it is the bond between Babylon and Las Vegas or Nineveh and Monterrey. It may be that urbanisation and industrial development is the key to the future of the arid zone as well as an element of its past.

Communities of whatever size in the arid lands depend on routeways and the larger the community the more efficient the communication links must be. Modern desert transport is a far cry from the days when Sir David Baird marched across the desert between Kosseir (Quseir) on the Red Sea to Kuft (Qus) in 1801 losing three men out of 5,000, many more nearly dying through the rottenness of the water-skins; yet by 1839 this same route between Kosseir and Kenneh (Qena) was being used by the

Peninsular and Orient Line to carry passengers and mails in transhipment to the Nile to avoid sailing right up the Gulf of Suez. In the Egyptian desert west of the Nile began those early experiments with motor vehicles for desert transport followed by the French Citroën expedition which in 1922 took twenty days for the crossing from Algiers to Timbuktu using half-track vehicles. Yet by 1942–3 General Leclerc was able to move a major military force from south to north across the Sahara, widening the pass of Kourizo, the entry into the Tibesti, with dynamite. Today the legend is told of the traveller for a well-known apéritif who arrived at Zouar in Chad from the Fezzan not knowing that the Jeep he was driving had four-wheeled drive, while the American tourist is provided with desert buggies and scooters to help him seek, but never find, the legendary lost gold mines of California, Nevada and New Mexico. On 12 February 1967, twelve sand yachts set out from Colomb Béchar to race behind a Land Rover to Nouakchott in Mauritania via Tindouf, Zouerat and Cap Timiris. The 3,218 km (2,000 miles) journey was completed thirty-one days later, an achievement which would appear to indicate that the ship of the desert is a reality rather than simply a synonym for the camel.

Yet it is with industry and the exploitation of minerals that one sees most impressive use of transport in arid regions. The mining companies of the Chilean desert have constructed two motorways to cope with the increase in motorised traffic while, in North Africa by 1962, two motorways converged on Hassi–Messaoud, one from Algiers by Hassi R'Mel and the other from Algiers by Biskra and Touggourt. The islands of oil exploitation in the Sahara are accompanied by airstrips and lorry transport and it is now commonplace to see convoys of ten-ton lorries carrying heavy machinery parked in the same layby as a camel caravan by the water-holes of the desert tracks. Beyond the oil-fields these lorries now make the journey from Algiers to Tamanrasset in a six-week round-trip with few precautions other than travelling in convoy through politically unstable areas, which lie north of the desert proper, following a definite itinerary and keeping ample reserves of food, water and petrol. It has been estimated that motorised transport, using the four main tracks across the Sahara, is already carrying each year a thousand passengers and over 2,000 tons of merchandise, principally groundnuts, henna, native butter and skins to the north and machinery, consumer goods and Algerian wine to the south. The lorry now takes four days for the journey between Taoudeni and Timbuktu instead of thirty days needed by the camel salt caravan, although it is believed that the cost per ton/mile is two to three times greater than by the traditional method. Even so the commercial camel caravan cannot survive much longer.

The exploitation of minerals in the arid lands was originally, with the exception of salt, limited to the discovery of precious metals since, with difficulties of transport and apparent lack of local energy supplies, only gold and silver could lure the prospector and mining speculator into the wilderness. Yet throughout the dry zone there are mineral deposits which are now being exploited or await exploitation. In Mongolia, silver, lead and coal are available; there is iron ore in the Gobi desert while 10 per cent of the world output of copper comes from Chuquimata in the dry zone of Chile, a country which also produces about three million tons of nitrates each year from its desert area. The town of Maria Elena, dating from 1926, houses two exploiting companies while the third is situated in another town, Pedro de Valdivia, specially built for the purpose about twenty miles away. To them must be brought all supplies of food and water since none is available locally. In Western Australia the discovery of gold reached its peak in August 1893 when at Coolgardie 500 ounces were taken from an exposed reef in a few hours. Here, too, water was a problem which was solved by the Government which first arranged for supply points along the trail and later pumped water from a reservoir at Mundaring near the coast 531 km (330 miles) to the east by pipeline. Gold, silver and borax brought men and towns to the American deserts, although the majority of the settlements are now often reduced to tourist attractions or serve as locations for television films.

For the extraction of these precious metals water problems existed as they did for agriculture and pastoralism. Each society in the arid lands has solved the problem of water rights by precedents and traditional practice. In Algeria, for instance, there are two main methods; under one system land and water are both private property and are dealt with as a unit for legal purposes with but few exceptions. Under the other system, which obtains more in the drier south, water is the only valuable property and is not tied to the land on which it is used—it may be sold or used as the group or individual determines. When the Mormons began to irrigate the land round Salt Lake City there were no problems of water ownership since Church and community were regarded as indivisible which made the establishment of communal water rights relatively easy. They had, moreover, been introduced to the Old World Mediterranean concepts of water rights through contacts with the Spanish settlers in New Mexico. There the system was based on the principle that a man could extract as much water from a river as was necessary for his purpose without the need to return any for the next user downstream. This was in direct conflict with the system derived from English common law where the riparian owner can expect to have water flowing through his land in undiminished quality and quantity—he can use the water as long as the same amount is returned to the stream at the lower end of his property. Such principles of riparian rights like so many other social, economic and legal principles exported from the humid to the arid zones, proved unworkable in the exploitation of placer mineral deposits. The miners of the alluvial gold in the dry USA adopted, through a system of precedents, a water rights system which was much more akin to the Spanish Mexican system, i.e. the right of appropriation; conflict between the farmers, pastoralists and miners was therefore inevitable. It was, moreover, aggravated by the different types of economic exploitation and especially by the greatly different financial returns. Even at the present day the principles of water rights are not completely codified and the systems vary between the different states.

If water was required in quantity for the extraction of precious metals in Colorado and California it is also a limiting factor in the exploitation of the ferrous and non-ferrous metals of the Sahara and for tapping the reserves of oil which lie deep beneath the surface. When the French struck oil at Hassi-Messaoud on the camel track between Touggourt and Fort Lallemand, it was fortunate that in the drilling they had already encountered vast quantities of water 2,133 m (7,000 ft) below the surface. That the water presented problems to the oil driller was, in the long run, less important than the fact that it also opened up 2,330 sq km (900 square miles) with assured water supplies in which forty-eight oil-producing wells have been drilled since the first successful strike on 15 June 1956. (Earlier, traces of gas had been discovered by drilling on the margins of Djebel Berga about 100 km [60 miles] south of In-Salah.) Now there are four major producing areas linked by pipelines and roads to the coast in this North African desert tract; three, Edjeleh-Zarzaitine, Hassi-Messaoud and Zelten (in Libya) produce oil, while at Hassi R'Mel both oil and natural gas are available. Together with the wells of the Middle East they have transformed the potential of the Afrasian desert tract by providing a direct competitor to agriculture and pastoralism with access to world rather than to local markets or mere subsistence economies.

Through the impetus of the discovery of oil in these Old World deserts there has been a transformation of attitudes towards the dry lands such as occurred in North America over a hundred years ago. The prospect of further mineral riches has enticed survey and drilling teams to prove the extent of the iron, lead, tin and copper which have been discovered and to verify reports of asbestos, nickel, platinum and uranium. Over a thousand miles to the south-west of the Saharan oil-fields in Mauritania there is already a new town, Zouerat, built to exploit the vast iron-ore reserves on the borders of Mauritania and the Rio D'Oro. A project of this magnitude involved extensive mechanisation and the building of a railway, with a capacity of 20,000 tons per day, 400 miles across difficult relief including an escarpment with an amplitude of 304 m (1,000 ft), apart from the dunes and sand-spreads, to an enlarged and developed terminal at Port Etienne. Elsewhere industrial oases such as Colomb Béchar are the complete antithesis of the traditional view of the North African deserts; Hassi-Messaoud in the oilfield is already planned to expand to a population of over 30,000 and will be complete with palm-lined streets and swimming pools using the abundant water from but 9 m (30 ft) below the surface. One might expect that experience with the arid climates of North Africa will persuade the planners to conform to the local environment rather than to import the architecture of the humid lands as has happened in the desert resorts of Reno and Las Vegas in the USA.

It has been suggested that the future of the arid lands lies in the extension of urban communities dependent on industry but with imported food supplies.[1] Where water is scarce it may be a more efficient use of land to establish an industrial township where, even with watering of gardens, 454 litres (100 gallons) per head per day would be less demanding of water than for purposes of stock rearing, or more especially for irrigation. It is pointed out that living conditions in a hot dry climate can be ideal with adequate supplies of water and that economies are possible in the use of air conditioning rather than winter heating. Other practical

[1] Sir L. Dudley Stamp, 'Urbanisation in Arid Lands' (in) Land Use in Semi-Arid Mediterranean climates (1964), UNESCO, Arid Zone Research XXVI, pp. 167–8.

problems, however, still remain such as the disposal of sewerage and industrial waste which demands vast quantities of water unless expensive chemical conversion plants are installed. There may be competition for water for domestic and industrial purposes or competition between neighbouring towns for available water as in southern California. There Los Angeles has been able to price smaller communities out of the water market and can obtain its supplies from the Sierra Nevada and Owens Valley; as a result the smaller urban communities are now part of the Los Angeles complex which increased its dominance by being able to provide for water-using industries. Even so there is a limit to the possible expansion of the conurbation since water supplies are not unlimited.

Such problems have already been discerned in the third largest city of Mexico far remote from the other major centres of population in the semi-arid north towards the Texas border. The town of Monterrey, founded by the settlement of twelve Spanish families in 1596, had a chequered career, much influenced by natural calamities of floods and disease and by the raids of Mexican Indians, until in the 1890s it was linked by rail to the United States and to the capital of Mexico City, and with the port of Tampico on the Gulf. With communications established, and in spite of a hostile natural and social environment, Monterrey became the leading industrial centre of Mexico, using water piped across the desert, by building numerous electricity generating plants. Together with cheap natural gas from Texas and the development of new iron and steel processes the town has now become the centre of iron and steel production in Mexico. New industries, however, are low water consumers since the problems of water supply are calling a halt to continued expansion on the earlier scale. To this environmental problem of Monterrey is also linked the social problems which arise from urbanisation in areas of poverty and subsistence economies.

Compared with its semi-arid hinterland, the prospects of high wages in Monterrey offer a great attraction to the rural populations and heavy rural emigration from the states of Nueva Leon and Coahuila to Monterrey has taken place. Only 32 per cent of the population of Nueva Leon works in agriculture, the lowest proportion of any of the Mexican states, and the proportion will continue to decrease as 5,000 people move in from the rural districts each year. The social problems to which this immigration gives rise have, of course, also been encountered in other towns in the arid lands, especially those where industry is by no means as well developed as in Monterrey. For many developing countries whose fortunes have been created by the discovery of oil, which even with local refineries is not a high labour-demanding industry, there has been an excessive development of the service industries (sometimes described as tertiary activities) in relation to the primary activity of oil production without the introduction of the secondary sector of other industrial employment, or the improvement of agriculture by irrigation or alternative crops and methods.

Many towns in North Africa and the Middle East show this trend such as Algiers, Oran and Tunis, but the most recent examples are in Libya where it has been remarked that everyone will be living in the towns with nobody to exploit the undoubted resources of the countryside as shown both by the plethora of Greek and Roman remains and, more recently, by the Italian colonisation schemes. Sheep, wheat, timber, fruit and vegetables, and wine (were not the Libyans followers of the Moslem faith) could be developed, since there are ample water supplies available from permanent springs which can be tapped by pipeline and from underground. Yet, as from the rural districts of Nueva Leon, there is a constant drift from the land to cities on the coast such as Benghazi and Tripoli and to the new capital of Beida in Tripolitania. Shanty towns spring up round the margins of the great office blocks and skyscraper flats, where in the most desirable areas land may cost as much as £70 a square metre. In the urban shanty slums live increasing numbers of landless labourers whose numbers increase as migration from the land accelerates in years of drought. Many of the smaller oases, especially in the Fezzan, now have a completely unbalanced age and sex structure as the young men have moved away leaving the women and the aged to carry on the traditional agricultural activities. The problem in Libya, as elsewhere in the Old World, is aggravated by the change from the nomadic way of life to sedentary activities although the nomad seems to have proved himself less adaptable to the problems of acquiring manual and technical skills than the sedentary cultivator. It is clear that, even with the great wealth available from Libya's natural resources, its Development Plan, to which was allocated 70 per cent of all oil revenues, may founder on urbanisation without the industrial activities and strong rural life on which real economic and social development must depend.

At Ouargla in the Algerian Sahara the provision of water from

artesian sources 1,219 m (4,000 ft) below the surface has assisted
in the revival of an old-established oasis where the palm trees
were dying; now new date palms are being planted and ground
given over to vegetables and cereals where formerly the flocks of
the pastoral nomads used to roam. Here, too, there are problems
of social integration of the nomad with the sedentary agriculturalist—problems which arise from the economic revolution
created by the application of new techniques to the exploitation
of old resources. This problem was also encountered in a more
drastic fashion in the richest sheikdom in the world, Kuwait,
which formerly derived a meagre income from customs dues
exacted on the shores of the Persian Gulf. From the original
settlement at the beginning of the eighteenth century until the
early 1950s, the people of Kuwait subsisted at a level comparable
with the desert Bedouin from which they had originally come.
Yet, almost overnight, the annual revenue rose to £50–60 million
per year derived from the oil beneath the sands. This great wealth
was superimposed on a feudal social structure composed of the
ruling family, the merchants and the people, in which the power
of the ruler is absolute and the merchants, dependent on the sea-
faring ability of the people, had built up wealth which maintained
a nice balance of power in the social organisation. Since the
revenue from the oil is vested in the ruler the economic power in
the community has changed and it was fortunate that he decided
to use the new wealth for the material, but to a larger extent, the
social transformation of his country. The rows of low mud houses
have been replaced by flats and houses, the narrow streets and
paths by ring roads and one-way streets; but there is also universal education, especially of the technical kind, and with the
development of industry has come employment for the people of
the sheikdom who already obtain much of their water from the
four-stage flash distillation plant fuelled by natural gas for the
desalination of sea-water. There is thus in Kuwait an awareness
of the problems which economic wealth can bring unless the
society and its attitudes are kept in step. Even in Kuwait, however,
there are difficulties which will intensify as new social groups
emerge, some antagonistic to the old form of society and others
antagonistic to the other social groups. Such difficulties would be
magnified for the whole of the oil-producing states in the Old
World arid lands with the exhaustion of the oil supplies (what
would happen to Kuwait's water supply?), or, in the more im-
mediate future, the supply of oil from sources less distant from

Europe such as the North Sea. The use of a new source of energy such as atomic power offers similar hazardous consequences to oil-producing countries in the Sahara and the Middle East. It may be that the only form of insurance against economic and social catastrophe is the investment of oil revenues into the traditional forms of land-use to provide new equipment for agriculture in the riverine lands and oases, for improvements in stock and for food-processing industries.

Diversification of economic effort is now well demonstrated both in the dry zones of North America and in Israel. Agriculture, pastoralism, industry, urbanisation and recreation, including tourism, have developed side by side and in symbiosis using to the best advantage the opportunities which the arid lands provide. With constant research into new energy resources, irrigation and desalination techniques, they are probably the keys to future arid land development, but they must take place in the correct social framework in which the elements of landlordism which have for so long bedevilled economic development in the arid lands of the Old World are solved. All elements of the community must have a sense of purpose, whether it is the opportunity for all according to his ability to increase his wealth, as in the USA, or a purpose which is superimposed from above, as in the USSR, or common interest in the maintenance of religion and culture in face of past oppression and current political problems.

For this last one might take the example of Israel in which the establishment of a new state in 1948 was accompanied by major immigration of Jews, not only from the surrounding countries but also from Europe so that in the first ten years of its existence the population doubled. The immigrants brought few skills of dry-land utilisation since they came either from the backward countries of the Middle East or from the humid lands of Europe where many had been trained in professions which had little relevance to, and practical acquaintance with, primary production. They came to the arid coastal strip of the eastern Mediterranean in an area of low rainfall with permeable rocks and sands margined by the barren wilderness of the Negev. There were, however, important compensations in that the land was by no means as desolate as other parts of the Afrasian desert tracts, there were technical skills available and aid in plenty, not only in expertise from other countries but also finance from the contributions of world Jewry and grants-in-aid from the American Government together with reparations from the German Government. It has been estimated that for each new family to be settled there was available 15,000 dollars, but a proportion of this money had to be spent in administration, defence, education and the improvement of the infrastructure, so that a major problem was the development of agriculture and industry in the face of an adverse balance of payments position—in comparison, for instance, with the very favourable export-import balances of the oil countries in the Old World arid lands. Investment in Israel and increased productivity in agriculture and industry improved the country's financial position but, in the context of arid land development, it was the large amount (about 40 per cent) invested in the primary activities, agriculture and industry, which is interesting.

The land under irrigation was increased threefold in the first ten years, by which time the country had reached self-sufficiency in eggs, poultry, milk, potatoes and fruits. Groundnuts, cotton and vegetables were exported, but overshadowed by the value and quantity of citrus fruits, under carefully controlled conditions of irrigation, where water is piped from the more humid areas in the north directly to the arid fields of the south. There can be few areas of the arid zone where irrigation water is used more efficiently. With the exception of cotton and food and minerals Israeli industry is primarily based on imported raw materials such as iron and steel which, although small units compared with those of Europe and the USA, supply the majority of local needs. Yet Israel is also aware that it has available the best opportunities for fertiliser production between the Mediterranean and the Pacific Ocean for which have been developed the growing chemical industries based on the great reserves of potash and bromine from the Dead Sea. Phosphate deposits are also available and the energy resource, oil, although imported largely from Persia, is piped from the developing port of Eilat at the head of the Gulf of Aqaba. Through this port now passes about 20 per cent of the country's exports, especially fertilisers and cement to East Africa, India and Japan. Into it came, in 1967, about 90 per cent of the oil requirements of the country, oil which is pumped directly to the port of Haifa on the Mediterranean coast. Into Eilat also come oilseeds from East Africa, to be processed into edible oils for transmission to Europe, together with rice, copra and timber. With a major tourist industry also, Eilat has become a major growth point of the country with a population of over 13,000 to whom special tax concessions have been an incentive to settle.

In its use of its eastern and western seaboards in relation to world trade, Israel is maintaining the age-old function of the Mediterranean coastland as a zone of trade and transhipment and exploiting its resources and geographical position in spite of the adverse environment. Yet the main lesson to be learnt from Israel in relation to the development of the arid lands is the emphasis which was placed on solving the educational and social problems of life in an arid environment. There has been emphasis on the need for motivation to the agricultural way of life and the acceptance of the limitations which this imposes on social life and economic returns. The public has been made aware of the basic unit of manpower, the family, and its relation to food production and consumption, and of the larger groupings of the village cooperatives, in which managerial qualities are required if the agricultural units of production, whether individual farm or village, are to be viable in a competitive internal and external market situation. Such desiderata have not always been achieved without major difficulties of adjustment and internal tensions and there has been the need for constant changes in the methods of introducing the new settlers to agriculture. Under any environmental conditions these would be difficult problems, even without external political pressures, although the latter may, in fact, be the key to their success.

It would appear, therefore, from the example of Israel that the application of science and technology to the utilisation of the dry lands is only one element of their successful occupation of the arid zone. The use of solar energy, the desalination of ground and sea-water, the creation of new architecture and living conditions to suit hot dry environments, the provision of new crops and better breeds of stock, the elaboration of agricultural and pastoral techniques, the reduction of diseases and pests, the surveys of soil, climate, vegetation, minerals and water supplies—these do not in themselves ensure economically successful and continuing occupancy of the dry lands of the world. To them must be added the creation of the right attitude towards life in the hot dry lands through education, social reform, and sound political organisation. Once these are grafted on to the achievements of science then perhaps all the arid world will achieve the living standards of California or Israel.

REFERENCES AND SELECTED BIBLIOGRAPHY

ADDISON, H., 1959, *Sun and Shadow at Aswan*, London
—1961, *Land, Water and Food*, London

AHMAD, K. S., 1951, 'Climatic Regions of West Pakistan', *Pakistan Geogr. Rev.*, *6*, 1–35

AL-KHASHAB, W. H., 1958, 'The water budget of the Tigris and Euphrates basin', Dept. of Geography, Univ. of Chicago. *Research Paper No. 54*

AMIRAN, D. H. K., 1954, 'The geography of the Negev and the southern limit of settlement in Israel', *Israel Expl. Journ.*, *4*

ANTEVS, E., 1954, 'Climate of New Mexico during the last glacio-pluvial', *J. Geol.*, *62*, 182–91

ARBOS, P., 1923, 'The geography of pastoral life', *Geogr. Rev.*, *13*, 559–75

AWAD, M., 1954, 'The assimilation of Nomads in Egypt', *Geogr. Rev.*, *44*, 240–52

BAGNOLD, R. A., 1941, *The Physics of Blown Sand and Desert Dunes*, London

BAGNOULS, F., 1957, 'Les climats biologiques et leur classification', *Annls. Géogr.*, *66*, 193–220

BAGOT-GLUBB, Sir J., 1960, *War in the Desert*, London

BARBOUR, K. M., 1959, 'Irrigation in the Sudan', *Trans. Inst. Br. Geogr.*, *26*, 243–63

BARTH, F., 1960, 'The land use pattern of migratory tribes of southern Persia', *Norsk. G. Tids.*, *17*, 1–11

BEAUJEU-GARNIER, J., 1955, 'Les oasis sahariennes', *Géographia*, *44*, 8–15

BILLINGTON, R. A., 1960, *Westward Expansion*, London

BIROT, P. and DRESCH, J., 1953, *La Mediterraneé et le moyen Orient*, 2 vols., Paris

BJERRE, J., 1960, *Kalahari*, London

BLACHE, J., 1921, 'Modes of life in the Moroccan countryside', *Geogr. Rev.*, *11*, 477–502

BLACKWELDER, E., 1931, 'Rock cut surfaces in desert ranges', *J. Geol.*, *20*, 442–50

BLUM, H. F., 1945, 'The physiological effects of sunlight on man', *Physiol. Rev.*, *25*, 483–530

BOSAZZA, V. L., 1954, 'Problems of water supply in the arid areas', *Geogr. J.*, *120*, 119–22

BOVILL, E. W., 1933, *Caravans of the Old Sahara*, London
—1958, *The Golden Trade of the Moors*, London

BOWMAN, I., 1924, *Desert Trails of Atacama*, Am. Geogr. Soc., New York
—1935, 'Our expanding and contracting desert', *Geogr. Rev.*, *25*, 43–61

BRICE, W. C., 1954, 'Caravan traffic across Asia', *Antiquity*, *28*, 78–84

BROWN, R. M., 1927, 'The utilisation of the Colorado river', *Geogr. Rev.*, *17*, 452–66

BRYAN, K., 1927, 'Persistence of features in an arid landscape', *Geogr. Rev.*, *17*, 251–57
—1935, 'The formation of pediments', *Rept. 10th Int. Geol. Cong.*, Pt. 2, 765–75
—1940, 'The retreat of slopes', *Ann. Ass. Am. Geog.*, *30*, 254–68

BUTZER, K. W., 1964, *Environment and Archaeology*, London

CALDER, R., 1951, *Men Against the Desert*, London

CAPOT-REY, R., 1953, *Le Sahara Français*, Paris

CAREY, P. C. and A. G., 1960, 'Oil and economic development in Iran', *Pol. Sci. Quart.*, *75*, 66–86

CHAPELLE, J., 1958, *Nomades noirs du Sahara*, Paris

CHAPMAN, V. J., 1960, *Salt Marshes and Salt Deserts of the World*, London

CHURCH, R. J. H., 1961, 'Problems and development of the dry zone of West Africa', *Geogr. J.*, *127*, 187–204

CLARK, J. I., 1959, 'Studies of semi-nomadism in north Africa', *Econ. Geogr.*, *35*, 95–108
—1963, 'Oil in Libya: some implications', *Econ. Geogr.*, *39*, 40–59

CLOUDSLEY-THOMSON, J. L. [Ed.], 1954, *Biology of Deserts*. Institute of Biology, London

CLOUDSLEY-THOMSON, J. L. [and CHADWICK, M. J.], 1964, *Life in Deserts*, London

CLOUDSLEY-THOMSON, J. L., 1965, *Desert Life*, London
—1960, *The Australian Environment*. 3rd ed. Revised. C.S.I.R.O., London and Melbourne

COTTON, C. A., 1942, *Climatic Accidents in Landscape Making*, Christchurch

CRARY, D. D., 1951, 'Recent agricultural developments in Saudi Arabia', *Geogr. Rev.*, *41*, 366–83

CRESSEY, G. B., 1957, 'Water in the Desert', *Ann. Ass. Am. Geog.*, *47*, 105–24
—1959, 'Deserts in Asia', *Proc. I.G.U. Reg. Cong. in Japan, 1957*, 109–12
—1960, *Crossroads: Land and Life in South-west Asia*, Philadelphia

CROWLEY, F. K., 1960, *Australia's Western Third*, London

DAVIS, W. M., 1905, 'The geographical cycle in an arid climate', *J. Geol.*, *13*, 381–407
—1931, 'Rock floors in arid and in humid climates', *J. Geol.*, *38*, 1–27, 136–57

DEBENHAM, F., 1953, *Kalahari Sand*, London

DICKSON, H. R. P., 1956, *Kuwait and Her Neighbours*, London

DOUGHTY, C. M., 1926, *Wanderings in Arabia*, London

DRESCH, J., 1966, 'Utilisation and human geography of the deserts', *Trans. Inst. Br. Geogr.*, *40*, 1–10

FARMER, B. H., 1954, 'Problems of land use in the dry zone of Ceylon', *Geogr. J.*, *120*, 21–33

FIELD, N. C., 1954, 'The Amu Darya: a study in resource geography', *Geogr. Rev.*, *44*, 528–42

FISHER, W. B., 1953, *The Middle East*, London

FULLER, M. C., 1924, 'Loess and rock dwellings of Shensi China', *Geogr. Rev.*, *14*, 215–26

GAITSKELL, A., 1959, *Gezira*, London

GAUTIER, G. F., 1923, *Le Sahara*, Paris
—1926, 'The Ahaggar: heart of the Sahara', *Geogr. Rev.*, *16*, 378–94

GLUECK, N., 1959, *Rivers in the Desert*, New York

GOETZMANN, W. H., 1959, *Army Exploration in the American West, 1803–1863*, Yale and London

GOTTMANN, J., 1938, 'L'homme, la routé et l'eau en Asie sud-occidentale', *Annls. Géogr.*, *47*, 575–601

GROVE, A. T., 1960, 'The geomorphology of the Tibesti region', *Geogr. J.*, *126*, 18–31

HAMMING, E., 1958, 'Water legislation', *Econ. Geogr.*, *34*, 42–46

HARRIS, W. B., 1897, 'The nomadic Berbers of Central Morocco', *Geogr. J.*, *9*, 633–45

HELLSTRÖM, B., 1953, 'The ground-water supply of north-eastern Sinai', *Geografiska Annaler*, *35*, 61–74

HILLS, E. S. [Ed.], 1966, *Arid Lands*, UNESCO, London

HOLM, D. A., 1960, 'Desert morphology in the Arabian peninsula', *Science*, *132*, 1369–79

HOLMES, C. D., 1955, 'Geomorphic development in humid and arid regions', *Am. J. Sci.*, *253*, 337–90

HOOVER, J. W., 1931, 'Navajo nomadism', *Geogr. Rev.*, *21*, 429–45

HOSTIE, J. F., 1955, 'Problems of international law concerning irrigation of arid lands', *International Affairs*, *31*, No. 1

HOUSTON, J. M., 1954, 'The significance of irrigation in Morocco's economic development', *Geogr. J.*, *120*, 314–28

HUNTINGTON, E., 1907, *The Pulse of Asia*, New York
—1914, *The Climatic Factor as Illustrated in Arid America*, New York

IVES, R. L., 1949, 'Climate of the Sonora desert region', *Ann. Ass. Am. Geog.*, *39*, 143–87

JAEGER, E. C., 1957, *The North American Deserts*, Stanford and London

JARVIS, C. S., 1938, *Desert and Delta*, London

JOHNSON, D. W., 1931, 'Plains of lateral corrasion', *Science*, *73*, 174–77
—1932, 'Rock planes in arid regions', *Geogr. Rev.*, *22*, 656–65

KANITKAR, N. V., 1960, *Dry Farming in India*, New Delhi

KEAST, A. [Ed.], 1959, 'Biogeography and ecology in Australia', Monographia Biologicae VIII, The Hague

KING, L. C., 1953, 'Canons of landscape evolution', *Bull. Geol. Soc. Am.*, *64*, 721–52

LAWRENCE, T. E., 1935, *The Seven Pillars of Wisdom*, London

LAWSON, A. C., 1915, 'The epigene profiles of the desert', *Univ. Calif. Bull.*, *No. 9*, 25–48

LEBON, J. H. G., 1955, 'The new irrigation era in Iraq', *Econ. Geogr.*, *31*, 47–59

LEOPOLD, L. B., 1951, 'Pleistocene climate in New Mexico', *Am. J. Sci.*, *249*, 152–67

LOGAN, R. F., 1960, *The Central Namib Desert, South-west Africa,* National Research Council [Publication 758], Washington, D.C.

LONGRIGG, S. H., 1961, *Oil in the Middle East,* London

LOWDERMILK, W. C., 1960, 'The reclamation of a man-made desert', *Scientific American, 202,* 55–63

LYDOLPH, P. E., 1951, 'A comparative analysis of the dry western littorals', *Ann. Ass. Am. Geog., 47,* 213–30

MAITLAND, L., 1960, *Forest Venture: Conquering the Deserts of the Middle East,* London

MARMER, H. A., 1951, 'The Peru and Niño Currents', *Geogr. Rev., 41,* 331–8

MARTIN, H., 1957, *The Sheltering Desert,* London

MARTONNE, E. de, 1926, 'Aréisme et indice d'aridité', *C. R. Acad. Sci.* [*de Paris*], *182,* 1395–98
—1927, 'Regions of interior basin drainage', *Geogr. Rev., 17,* 397–414

MCGEE, W. J., 1897, 'Sheetflood erosion', *Bull. Geol. Soc. Am., 8,* 87–112

MEIGS, P., 1952, 'Water problems in the USA', *Geogr. Rev., 42,* 346–66
—1953, 'Design and use of homoclimatic maps', *Proc. Int. Symp. Desert Research,* Jerusalem
—1966, *A Geography of Coastal Deserts,* Arid Zone Research, No. XXVIII, UNESCO, Paris

MEINZER, O. E., 1927, 'The occurrence of ground-water in the United States', *Water Supply Paper 489,* U.S. Geol. Surv., Washington, D.C.

MERRYLEE, J. K., 1959, 'Water problems in the Middle East', *J. Cent. Asian Soc., 46,* 39–45

MILLER, A. A., 1931, *Climatology,* London

MONOD, T., 1958, 'Majâbat al-Koubrâ', *Memoire de l'IFAN, 52,* 141–50

MORRIS, J., 1961, *Masters of the Desert. 6,000 Years in the Negev,* New York

MURDOCK, G. P., 1960, 'Staple subsistence crops in Africa', *Geogr. Rev., 50,* 523–40

MURRAY, G. W., 1955, 'Water from the desert: some ancient Egyptian achievements', *Geogr. J., 121,* 171–81

NICOLAISEN, J., 1954, 'Some aspects of the problem of nomadic cattle breeding among the Tuareg of the central Sahara', *G. Tids.* [*Copenhagen*], *53,* 62–105

NIXON, R. W., 1952, 'Ecological study of the date varieties in French North Africa', *Ecology, 33,* 215–25

OSBORN, F., 1954, *The Limits of the Earth,* London

PAVER, G. L., 1947, 'Water supply in the Middle East campaign', *Water and Water Engineering, 49,* 653-62

PEEL, R. F., 1960, 'Some aspects of desert geomorphology', *Geogr., 45,* 241-62

—1966, 'The landscape in aridity', *Trans. Inst. Br. Geogr., 38,* 1-23

PETROV, M. P., 1962, 'Types de déserts de l'Asie Centrale', *Annls. Géogr., 384,* 131-55

PHILBY, H. St. J. B., 1952, *Arabian Highlands,* London

POQUET, J., 1963, *Les Déserts,* 'Que Sais-Je?', No. 500, Paris

POWELL, J. W., 1878, 'Report of the Lands of the Arid Regions of the US with a More Detailed Account of the Lands of Utah', *45th Cong. 2nd Session, House Ex. Doc. 73,* Washington, D.C.

POWERS, W. C., 1954, 'Soil and land-use capabilities in Iraq', *Geogr. Rev., 44,* 373-80

PRENANT, A., 1953, 'Facteurs du peuplement d'une ville d'Algérie intérieure', *Annls. Geogr., 62,* 434-51

RAINEY, R. C., 1951, 'Weather and the movement of locust swarms: a new hypothesis', *Nature, 168,* 1057-60

REIFENBERG, A., 1955, *The Struggle Between the Desert and the Sown,* Jerusalem

RICH, J. L., 1935, 'Origin and evolution of rock fans and pediments', *Bull. Geol. Soc. Am., 46,* 999-1024

ROSS, C. G., 1960, 'Reducing water loss in South Australia', *Geogr., 45,* 297-99

ROY, J. M., 1954, 'La Grande Valleé de Californie', *Canadian Geogr., 4,* 63-76

RUDOLPH, W. E., 1927, 'The Ria Loa of Northern Chile', *Geogr. Rev., 17,* 553-85

—1951, 'Chuquicamata twenty years later', *Geogr. Rev., 41,* 88-113

RUSSELL, R. J., 1945, 'Climates of Texas', *Ann. Ass. Am. Geog., 35,* 37-52

SANGER, R. H., 1954, *The Arabian Peninsula,* Ithaca

SAUER, C. O., 1952, *Agricultural Origins and Dispersals,* Am. Geog. Soc., New York

SCHULZE, B. R., 1947, 'The climates of South Africa according to the classifications of Köppen and Thornthwaite', *S. Afr. geogr. J., 29,* 32-102

SEMPLE, E. C., 1931, 'Domestic and municipal waterworks in Ancient Mediterranean lands', *Geogr. Rev., 21,* 466-74

SHAPLEY, H. [Ed.], 1953, *Climatic Change. Evidence, Cause and Effects,* Cambridge, USA

SMITH, T. C., 1960, 'Aspects of agriculture and settlement in Peru', *Geogr. J.*, *126*, 397–412

SYKES, G., 1927, 'The Camino del Diablo', *Geogr. Rev.*, *17*, 62–74

SUBRAH MANYAM, U. P., 1956, 'The water balance of India', *Ann. Ass. Am. Geog.*, *46*, 300–11

SUSLOV, S. P., 1961, *Physical Geography of Asiatic Russia*, London

TAYLOR, G., 1918, *The Australian Environment*, Melbourne
—1939, 'Sea to Sahara—settlement zones in eastern Algeria', *Geogr. Rev.*, *29*, 177–95
—1940, *Australia*. 1st ed., London

THESIGER, W., 1959, *Arabian Sands*, London

THOMAS, B. E., 1957, 'Trade routes of Algeria and the Sahara', *Univ. of California Publications in Geography*, *8*, 165–288

THOMAS, W. L., 1959, 'Man, time and space in southern California', *Ann. Ass. Am. Geog.*, *49*, 1–120
—1960, 'Competition for a desert lake: the Salton sea, California', Abst. Papers XIX I.G.C., Norden

THORNTHWAITE, C. W., 1948, 'An approach towards a rational classification of climate', *Geogr. Rev.*, *38*, 55–94

TOTHILL, J. D., 1948, *Agriculture in the Sudan*, London

TRICART, J. and CAILLEUX, A., *Le modelé des régions seches*, Paris

UNESCO, Arid Zone Research, Paris, 1953—
Vol. I: Arid Zone Hydrology—Reviews of Research
—II: Arid Zone Hydrology—Proceedings of the Ankara Symposium
—III: Directory of Institutions Engaged in Arid Zone Research
—IV: Utilisation of Saline Water—Reviews of Research
—V: Plant Ecology—Proceedings of the Montpellier Symposium
—VI: Plant Ecology—Reviews of Research
—VII: Wind and Solar Energy—Proceedings of the New Delhi Symposium
—VIII: Human and Animal Ecology—Reviews of Research
—IX: Guide Book to Research Data for Arid Zone Development
—X: Climatology—Reviews of Research
—XI: Climatology and Microclimatology—Proceedings of the Canberra Symposium
—XII: Arid Zone Research—Recent Developments
—XIII: Medicinal Plants of the Arid Zones
—XIV: Salinity Problems in the Arid Zones
—XV: Plant-Water Relationships in Arid and Semi-Arid Conditions
—XVI: Plant-Water Relationships—Reviews of Research
—XVII: [Ed. Stamp, L. D.], A History of Land Use in Arid Regions
—XVIII: Problems of the Arid Zone—Proceedings of the Paris Symposium

—XIX: Nomades et Nomadisme au Sahara
—XX: Changes of Climate. Proceedings of the Rome Symposium
—XXI: Bioclimate Map of the Mediterranean Zone and Explanatory Note
—XXII: Environmental Physiology and Psychology in Arid Conditions—Reviews of Research
—XXIII: Agricultural Planning and Village Community in Israel
—XXIV: Environmental Physiology and Psychology in Arid Conditions
—XXV: Methodology of Plant Eco-Physiology—Proceedings of Montpellier Symposium
—XXVI: Land Use in Semi-Arid Mediterranean Climates
—XXVII: Evaporation Reduction
—XXVIII: Meigs, P., A Geography of Coastal Deserts

VERLET, B., 1962, *Le Sahara*, 'Que Sais-Je?', No. 766, Paris

WADHAM, S., 1957, *Land Utilisation in Australia*, Melbourne

WALTHER, J., 1924, *Das Gesetz der Wüstenbildung*, Leipzig

WALTON, K., 1952, 'The oasis of Jalo', *Scot. Geogr. Mag.*, *68*, 110–19

WALTON, K. [with GIMINGHAM, C. H.], 1954, Environment and the Structure of Scrub Communities on the Limestone Plateaux of Northern Cyrenaica, *Ecology*, *42*, 505–20

WAYLAND, E. J., 1953, 'More about the Kalahari', *Geogr. J.*, *119*, 49–56

WEULEKSSE, J., 1946, *Paysans de Syrie et du Proche-Orient* [Les paysans de la terre], Paris

WHITE, G. F. [Ed.], 1956, *The Future of Arid Lands*, Pubn. No. 43, Am. Ass. Adv. Sci., Washington

WHYTE, R. O., 1960, *Crop Production and Environment*, London

ZOHARY, M., 1962, *Plant Life of Palestine*, New York

INDEX

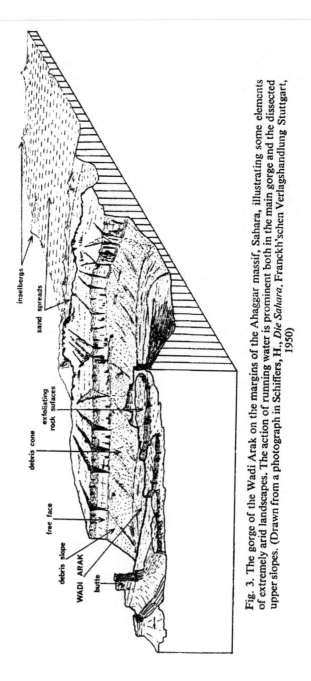

Fig. 3. The gorge of the Wadi Arak on the margins of the Ahaggar massif, Sahara, illustrating some elements of extremely arid landscapes. The action of running water is prominent both in the main gorge and the dissected upper slopes. (Drawn from a photograph in Schiffers, H., *Die Sahara*, Franckh'schen Verlagshandlung Stuttgart, 1950)

inselbergs

sand spreads

exfoliating rock surfaces

debris cone

free face

debris slope

WADI ARAK

butte